PIC BASIC Projects

PIC BASIC Projects

30 Projects Using PIC BASIC and PIC BASIC PRO

By

Dogan Ibrahim

ELSEVIER

AMSTERDAM • BOSTON • HEIDELBERG • LONDON • NEW YORK • OXFORD
PARIS • SAN DIEGO • SAN FRANCISCO • SINGAPORE • SYDNEY • TOKYO

Newnes is an imprint of Elsevier
Linacre House, Jordan Hill, Oxford OX2 8DP, UK
30 Corporate Drive, Suite 400, Burlington, MA 01803, USA

First edition 2006
Reprinted 2007

British Library Cataloguing in Publication Data
A catalogue record for this book is available from the British Library

Library of Congress Cataloging-in-Publication Data
A catalog record for this book is available from the Library of Congress

ISBN: 978-0-7506-6879-8

For information on all Newnes publications
visit our website at www.newnespress.com

Printed and bound by CPI Group (UK) Ltd, Croydon, CR0 4YY

Transferred to Digital Print 2012

Working together to grow
libraries in developing countries

www.elsevier.com | www.bookaid.org | www.sabre.org

ELSEVIER   BOOK AID
            International   Sabre Foundation

Trademarks/Registered Trademarks
PIC is a registered trademark of Microchip Technology Inc.

All brand names mentioned in this book are protected by their
respective trademarks and are acknowledged.

Contents

Preface

Microcontrollers are single-chip computers consisting of CPU (central processing unit), data and program memory, serial and parallel I/O (input/output), timers, external and internal interrupts, all integrated into a single chip that can be purchased for as little as $2.00. Microcontrollers are intelligent electronic devices used to control and monitor devices in the real world. Today microcontrollers are used in most commercial and industrial equipment. About 40% of microcontroller applications are in office automation, such as PCs, laser printers, fax machines, intelligent telephones, and so forth. About one-third of microcontrollers are found in consumer electronics goods. Products such as CD players, hi-f- equipment, video games, washing machines and cookers fall into this category. The communications market, automotive market, and the military share the rest of the application areas.

Microcontrollers are programmed devices. A program is a sequence of instructions that tell the microcontroller what to do. Microcontrollers have traditionally been programmed using the low-level assembly language of the target processor. This consists of a series of instructions in the form of mnemonics. The biggest disadvantage of assembly language is that microcontrollers from different manufacturers have different assembly languages and the user is forced to learn a new language every time a new processor is chosen. Assembly language is also difficult to work with, especially during the development, testing, and maintenance of complex projects. The solution to this problem has been to use a high-level language to program microcontrollers. A high-level language consists of easy to understand, more meaningful series of instructions. This approach makes the programs more readable and also portable. The same high-level language can usually be used to program different types of microcontrollers. Testing and the maintenance of microcontroller-based projects are also easier when high-level languages are used.

This book is about programming microcontrollers using a high-level language. The PIC family of microcontrollers is chosen as the target microcontroller. PIC is currently one of the most popular microcontrollers used by many engineers, technicians, students, and hobbyists. PIC microcontrollers are manufactured in different sizes and in varying complexity. These microcontrollers incorporate a RISC (reduced instruction set computer) architecture and there is only a small set of instructions that the user has to learn. Also, the power consumption of PIC microcontrollers is very low and this is one of the reasons which make these microcontrollers popular in portable hand-held applications.

In this book, PicBasic and PicBasic Pro languages are used to program PIC microcontrollers. BASIC is one of the oldest and widely known high-level programming languages. Both PicBasic and PicBasic Pro have been developed by MicroEngineering Labs Inc. PicBasic is a low-cost compiler and is aimed at the lower end of the market, mainly for students and the hobby market.

PicBasic Pro is more expensive and it is a sophisticated professional compiler with many extra features. This compiler is aimed for engineers and other professional users of PIC microcontrollers.

This book will help technicians, engineers, and to those who chose electronics as a hobby. No previous experience with microcontrollers is assumed, and the PIC family of microcontrollers is introduced in detail. The book is practical and is supplied with many working hardware projects where the reader can experiment easily using a simple breadboard type experiment kit and a few components. The circuit diagram, flow diagram, and the code for each project are given and explained in detail.

Chapter 1 provides a review of the basic architecture of microcontrollers. Various microcontroller concepts are described in this chapter.

Chapter 2 is about the common features of PIC microcontrollers and describes in detail the architecture of various types of commonly used PIC microcontrollers and their use in electronic devices.

A microcontroller-based system development requires both hardware and software development tools. Chapter 3 describes the various commercially available PIC microcontroller development tools and gives a brief overview of how they can be used in project development.

PicBasic and PicBasic Pro languages are discussed in detail in Chapter 4. A brief description of each statement is given with an example.

Finally, in Chapter 5, many tested and working projects are given. These projects are organized in increasing complexity and the reader is recommended to follow this chapter in the given order.

Dogan Ibrahim

1
Microcontroller systems

1.1   Introduction

In 1969, Bob Noyce and Gordon Moore set up the Intel Corporation to manufacture memory chips for the mainframe computer industry. Later in 1971, the first microprocessor chip 4040 was manufactured by Intel for a consortium of two Japanese companies. These chips were basically designed for a calculator named *Busicom* which was one of the first portable calculators. This was a very simple calculator which could only add and subtract numbers, 4 bits (a nibble) at a time. 4040 chip was so successful that it was soon followed by Intel's 8-bit 8008 microprocessor. This was a simple microprocessor with limited resources, poorly implemented interrupt mechanisms, and multiplexed address and data busses. The first really powerful 8-bit microprocessor appeared in early 1974 as the Intel 8080 chip. This microprocessor had separate address and data busses with 64 K byte of address space which was enormous in 1975 standards. 8080 microprocessor was the first microprocessor used in homes as a personal computer named *Altair*. 8080 has been a very successful microprocessor but soon other companies began producing microprocessor chips. Motorola introduced the 8-bit 6800 chip which had a different architecture to the 8080 but has also been very popular. In 1976, Zilog introduced the Z80 microprocessor which was much more advanced than the 8080. The instruction set of Z80 was downward compatible with the 8080 and this made Z80 to be one of the most successful microprocessors of the time. Z80 was used in many microprocessor-based applications, including home computers and games consoles. In 1976, Motorola created a microprocessor chip called 6801 which replaced a 6800 chip plus some of the chips required to make a complete computer system. This was a major step in the evolution of the microcontrollers which are basically computers consisting of only one chip. In later years, we see many other microcontroller chips in the market, such as Intel 8048, 8049, 8051, Motorola 6809, Atmel 89C51, etc.

The term microcomputer is used to describe a system that includes a minimum of a microprocessor, program memory, data memory, and input–output (I/O). Some microcomputer systems include additional components such as timers, counters, analogue-to-digital converters, and so on. Thus, a microcomputer system can be anything from a large computer having hard disks, floppy disks, and printers, to a single-chip embedded controller.

In this book we are going to consider only the type of microcomputers that consists of a single silicon chip. Such microcomputer systems are also called microcontrollers and they are used in many household goods such as microwave ovens, TV remote control units, cookers, hi-fi equipment, CD players, personal computers, fridges, etc.

1.2   Microcontroller systems

A microcontroller is a single chip computer (see Figure 1.1). *Micro* suggests that the device is small, and *controller* suggests that the device can be used in control applications. Another term used for microcontrollers is *embedded controller*, since most of the microcontrollers are built into (or embedded in) the devices they control.

A microprocessor differs from a microcontroller in many ways. The main difference is that a microprocessor requires several other components for its operation, such as program memory and data memory, I/O devices, and external clock circuit. A microcontroller on the other hand has all the support chips incorporated inside the same chip. All microcontrollers operate on a set of instructions (or the user program) stored in their memory. A microcontroller fetches the instructions from its program memory one by one, decodes these instructions, and then carries out the required operations.

Microcontrollers have traditionally been programmed using the assembly language of the target device. Although the assembly language is fast, it has several disadvantages. An assembly program consists of mnemonics and it is difficult to learn and maintain a program written using the assembly language. Also, microcontrollers manufactured by different firms have different assembly languages and the user is required to learn a new language every time a new microcontroller is used. Microcontrollers can also be programmed using a high-level language, such as BASIC, PASCAL, and C. High-level languages have the advantage that it is much easier to learn a high-level language than the assembler. Also, very large and complex programs can easily be developed using a high-level language. In this book we shall be learning the programming of PIC microcontrollers using the popular PicBasic and PicBasic Pro compilers.

In general, a single chip is all that is required to have a running microcontroller system. In practical applications additional components may be required to allow a microcomputer to interface to its environment. With the advent of the PIC family of microcontrollers the development time of an electronic project has reduced to several hours. Developing a PIC microcontroller-based project simply takes no more than five or six steps.

1. Type the program into a PC
2. Assemble (or compile) the program
3. Optionally simulate the program on a PC
4. Load the program into PIC's program memory
5. Design and construct the hardware
6. Test the project.

Basically, a microcomputer executes a user program which is loaded in its program memory. Under the control of this program data is received from external devices (inputs), manipulated and then sent to external devices (outputs). For example, in a microcontroller-based oven temperature control system the temperature is read by the microcomputer using a temperature sensor. The microcomputer then operates a heater or a fan to control and keep the temperature at the required value. Figure 1.2 shows the block diagram of our simple oven temperature control system.

Figure 1.1   Some PIC microcontrollers

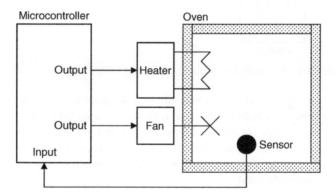

Figure 1.2   Microcontroller-based oven temperature control system

The system shown in Figure 1.2 is a very simplified temperature control system. In a more sophisticated system we may have a keypad to set the temperature, and a liquid crystal display (LCD) to display the current temperature. Figure 1.3 shows the block diagram of this more sophisticated temperature control system.

We can make our design even more sophisticated (see Figure 1.4) by adding an audible alarm to inform us if the temperature is outside the required values. Also, the temperature readings can be sent to a PC every second for archiving and further processing. For example, a graph of the daily temperature can be plotted on the PC. As you can see, because the microcontrollers are programmable it is very easy to make the final system as simple or as complicated as we like.

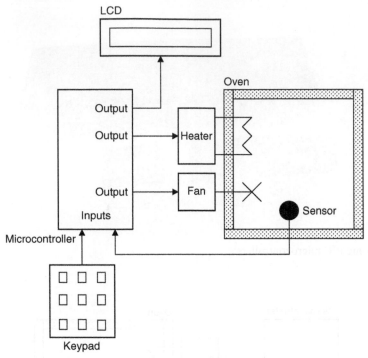

Figure 1.3   Temperature control system with a keypad and LCD

A microcontroller is a very powerful tool that allows a designer to create sophisticated I/O data manipulation under program control. Microcontrollers are classified by the number of bits they process. 8-bit microcontrollers are the most popular ones and are used in most microcontroller-based applications; 16- and 32-bit microcontrollers are much more powerful, but usually more expensive and not required in many small- to medium-size general-purpose applications where microcontrollers are generally used.

As shown in Figure 1.5, the simplest microcontroller architecture consists of a microprocessor, memory, and I/O. The microprocessor consists of a central processing unit (CPU) and the control unit (CU). The CPU is the brain of the microcontroller and this is where all of the arithmetic and logic operations are performed. The CU controls the internal operations of the microprocessor and sends out control signals to other parts of the microcontroller to carry out the required instructions.

Memory is an important part of a microcontroller system. Depending upon the type used we can classify memories into two groups: program memory and data memory. Program memory stores the program written by the programmer and this memory is usually non-volatile, i.e. data is not lost after the removal of power. Data memory is where the temporary data used in a program are stored and this memory is usually volatile, i.e. data is lost after the removal of power.

There are basically five types of memories as summarised below.

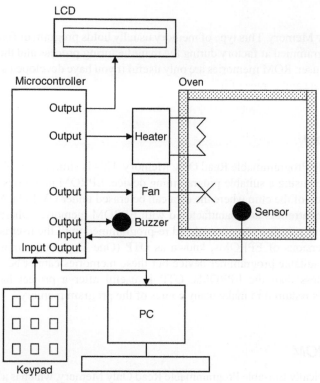

Figure 1.4   More sophisticated temperature controller

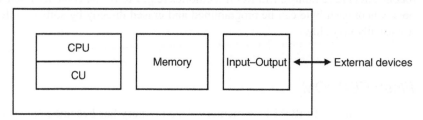

Figure 1.5   The simplest microcontroller architecture

1.2.1   RAM

RAM means Random Access Memory. It is a general-purpose memory which usually stores the user data used in a program. RAM is volatile, i.e. data is lost after the removal of power. Most microcontrollers have some amount of internal RAM. 256 bytes is a common amount, although some microcontrollers have more, some less. In general it is possible to extend the memory by adding external memory chips.

1.2.2   ROM

ROM is Read Only Memory. This type of memory usually holds program or fixed user data. ROM memories are programmed at factory during the manufacturing process and their contents cannot be changed by the user. ROM memories are only useful if you have developed a program and wish to order several thousand copies of it.

1.2.3   EPROM

EPROM is erasable Programmable Read Only Memory. This is similar to ROM, but the EPROM can be programmed using a suitable programming device. EPROM memories have a small clear glass window on top of the chip where the data can be erased under UV light. Many development versions of microcontrollers are manufactured with EPROM memories where the user program can be stored. These memories are erased and re-programmed until the user is satisfied with the program. Some versions of EPROMs, known as OTP (One Time Programmable), can be programmed using a suitable programmer device but these memories cannot be erased. OTP memories cost much less than the EPROMs. OTP is useful after a project has been developed completely and it is required to make many copies of the program memory.

1.2.4   EEPROM

EEPROM is Electrically Erasable Programmable Read Only Memory, which is a non-volatile memory. These memories can be erased and also be programmed under program control. EEPROMs are used to save configuration information, maximum and minimum values, identification data, etc. Some microcontrollers have built-in EEPROM memories (e.g. PIC16F84 contains a 64-byte EEPROM memory where each byte can be programmed and erased directly by software). EEPROM memories are usually very slow.

1.2.5   Flash EEPROM

This is another version of EEPROM-type memory. This memory has become popular in microcontroller applications and is used to store the user program. Flash EEPROM is non-volatile and is usually very fast. The data is erased and then re-programmed using a programming device. The entire contents of the memory should be erased and then re-programmed.

1.3   Microcontroller features

Microcontrollers from different manufacturers have different architectures and different capabilities. Some may suit a particular application while others may be totally unsuitable for the same application. The hardware features of microcontrollers in general are described in this section.

1.3.1 Supply voltage

Most microcontrollers operate with the standard logic voltage of $+5$ V. Some microcontrollers can operate at as low as $+2.7$ V and some will tolerate $+6$ V without any problems. You should check the manufacturers' data sheets about the allowed limits of the power supply voltage.

A voltage regulator circuit is usually used to obtain the required power supply voltage when the device is to be operated from a mains adaptor or batteries. For example, a 5 V regulator is required if the microcontroller is to be operated from a 5 V supply using a 9 V battery.

1.3.2 The clock

All microcontrollers require a clock (or an oscillator) to operate. The clock is usually provided by connecting external timing devices to the microcontroller. Most microcontrollers will generate clock signals when a crystal and two small capacitors are connected. Some will operate with resonators or external resistor–capacitor pair. Some microcontrollers have built-in timing circuits and they do not require any external timing components. If your application is not time-sensitive you should use external or internal (if available) resistor–capacitor timing components for simplicity and low cost.

An instruction is executed by fetching it from the memory and then decoding it. This usually takes several clock cycles and is known as the *instruction cycle*. In PIC microcontrollers an instruction cycle takes four-clock periods. Thus, the microcontroller is actually operated at a clock rate which is a quarter of the actual oscillator frequency.

1.3.3 Timers

Timers are important parts of any microcontroller. A timer is basically a counter which is driven either from an external clock pulse or from the internal oscillator of the microcontroller. A timer can be 8-bits or 16-bits wide. Data can be loaded into a timer under program control and the timer can be stopped or started by program control. Most timers can be configured to generate an interrupt when they reach a certain count (usually when they overflow). The interrupt can be used by the user program to carry out accurate-timing-related operations inside the microcontroller.

Some microcontrollers offer capture and compare facilities where a timer value can be read when an external event occurs, or the timer value can be compared to a preset value and an interrupt can be generated when this value is reached.

It is typical to have at least one timer in every microcontroller. Some microcontrollers may have two, three, or even more timers where some of the timers can be cascaded for longer counts.

1.3.4 Watchdog

Most microcontrollers have at least one watchdog facility. The watchdog is basically a timer which is refreshed by the user program and a reset occurs if the program fails to refresh the watchdog. The watchdog timer is used to detect a system problem, such as the program being in an endless loop. A watchdog is a safety feature that prevents runaway software and stops the microcontroller from executing meaningless and unwanted code. Watchdog facilities are commonly used in real-time systems where it is required to regularly check the successful termination of one or more activities.

1.3.5   Reset input

A reset input is used to reset a microcontroller. Resetting puts the microcontroller into a known state such that the program execution starts from address 0 of the program memory. An external reset action is usually achieved by connecting a push-button switch to the reset input such that the microcontroller can be reset when the switch is pressed.

1.3.6   Interrupts

Interrupts are very important concepts in microcontrollers. An interrupt causes the microcontroller to respond to external and internal (e.g. a timer) events very quickly. When an interrupt occurs the microcontroller leaves its normal flow of program execution and jumps to a special part of the program, known as the *Interrupt Service Routine* (ISR). The program code inside the ISR is executed and upon return from the ISR the program resumes its normal flow of execution.

The ISR starts from a fixed address of the program memory. This address is also known as the *interrupt vector address*. For example, in a PIC16F84 microcontroller the ISR starting address is 4 in the program memory. Some microcontrollers with multi-interrupt features have just one interrupt vector address, while some others have unique interrupt vector addresses, one for each interrupt source. Interrupts can be nested such that a new interrupt can suspend the execution of another interrupt. Another important feature of a microcontroller with multi-interrupt capability is that different interrupt sources can be given different levels of priority.

1.3.7   Brown-out detector

Brown-out detectors are also common in many microcontrollers and they reset a microcontroller if the supply voltage falls below a nominal value. Brown-out detectors are safety features and they can be employed to prevent unpredictable operation at low voltages, especially to protect the contents of EEPROM-type memories.

1.3.8   Analogue-to-digital converter

An analogue-to-digital converter (A/D) is used to convert an analogue signal such as voltage to a digital form so that it can be read by a microcontroller. Some microcontrollers have built-in A/D converters. It is also possible to connect an external A/D converter to any type of microcontroller. A/D converters are usually 8-bits, having 256 quantisation levels. Some microcontrollers have 10-bit A/D converters with 1024 quantisation levels. Most PIC microcontrollers with A/D features have multiplexed A/D converters where more than one analogue input channel is provided.

The A/D conversion process must be started by the user program and it may take several hundreds of microseconds for a conversion to complete. A/D converters usually generate interrupts when a conversion is complete so that the user program can read the converted data quickly.

A/D converters are very useful in control and monitoring applications since most sensors (e.g. temperature sensor, pressure sensor, force sensor, etc.) produce analogue output voltages.

1.3.9   Serial I/O

Serial communication (also called RS232 communication) enables a microcontroller to be connected to another microcontroller or to a PC using a serial cable. Some microcontrollers have built-in hardware called USART (Universal Synchronous–Asynchronous Receiver–Transmitter) to implement a serial communication interface. The baud rate and the data format can usually be selected by the user program. If any serial I/O hardware is not provided, it is easy to develop software to implement serial data communication using any I/O pin of a microcontroller. We shall see in Chapter 4 how to use the PicBasic and PicBasic Pro statements to send and receive serial data from any pin of a PIC microcontroller.

Some microcontrollers incorporate SPI (Serial Peripheral Interface) or I²C (Integrated Inter Connect) hardware bus interfaces. These enable a microcontroller to interface to other compatible devices easily.

1.3.10   EEPROM data memory

EEPROM type data memory is also very common in many microcontrollers. The advantage of an EEPROM memory is that the programmer can store non-volatile data in such a memory, and can also change this data whenever required. For example, in a temperature monitoring application the maximum and the minimum temperature readings can be stored in an EEPROM memory. Then, if the power supply is removed for whatever reason, the values of the latest readings will still be available in the EEPROM memory.

PicBasic and PicBasic Pro languages provide special instructions for reading and writing to the EEPROM memory of a microcontroller which has such memory built-in.

Some microcontrollers have no built-in EEPROM memory, some provide only 16 bytes of EEPROM memory, while some others may have as much as 256 bytes of EEPROM memories.

1.3.11   LCD drivers

LCD drivers enable a microcontroller to be connected to an external LCD display directly. These drivers are not common since most of the functions provided by them can be implemented in software.

1.3.12   Analogue comparator

Analogue comparators are used where it is required to compare two analogue voltages. Although these circuits are implemented in most high-end PIC microcontrollers they are not common in other microcontrollers.

1.3.13   Real-time clock

Real-time clock enables a microcontroller to have absolute date and time information continuously. Built-in real-time clocks are not common in most microcontrollers since they can easily be implemented by either using a dedicated real-time clock chip, or by writing a program.

1.3.14   Sleep mode

Some microcontrollers (e.g. PIC) offer built-in sleep modes where executing this instruction puts the microcontroller into a mode where the internal oscillator is stopped and the power consumption is reduced to an extremely low level. The main reason of using the sleep mode is to conserve the battery power when the microcontroller is not doing anything useful. The microcontroller usually wakes up from the sleep mode by external reset or by a watchdog time-out.

1.3.15   Power-on reset

Some microcontrollers (e.g. PIC) have built-in power-on reset circuits which keep the microcontroller in reset state until all the internal circuitry has been initialised. This feature is very useful as it starts the microcontroller from a known state on power-up. An external reset can also be provided where the microcontroller can be reset when an external button is pressed.

1.3.16   Low power operation

Low power operation is especially important in portable applications where the microcontroller-based equipment is operated from batteries. Some microcontrollers (e.g. PIC) can operate with less than 2 mA with 5 V supply, and around 15 μA at 3 V supply. Some other microcontrollers, especially microprocessor-based systems where there could be several chips may consume several hundred milliamperes or even more.

1.3.17   Current sink/source capability

This is important if the microcontroller is to be connected to an external device which may draw large current for its operation. PIC microcontrollers can source and sink 25 mA of current from each output port pin. This current is usually sufficient to drive LEDs, small lamps, buzzers, small relays, etc. The current capability can be increased by connecting external transistor switching circuits or relays to the output port pins.

1.4   Microcontroller architectures

Usually two types of architectures are used in microcontrollers (see Figure 1.6): *Von Neumann* architecture and *Harvard* architecture. Von Neumann architecture is used by a large percentage of microcontrollers and here all memory space is on the same bus and instruction and data use the same bus. In the Harvard architecture (used by the PIC microcontrollers), code and data are on separate busses and this allows the code and data to be fetched simultaneously, resulting in an improved performance.

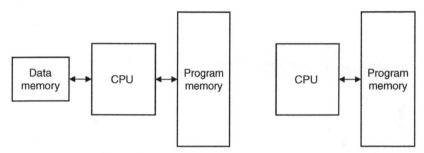

Figure 1.6   Von Neumann and Harvard architectures

1.4.1   RISC and CISC

RISC (Reduced Instruction Set Computer) and CISC (Complex Instruction Computer) refer to the instruction set of a microcontroller. In an 8-bit RISC microcontroller, data is 8-bits wide but the instruction words are more than 8-bits wide (usually 12, 14, or 16-bits) and the instructions occupy one word in the program memory. Thus, the instructions are fetched and executed in one cycle, resulting in an improved performance. PIC microcontrollers are RISC-based devices and they have no more than 35 instructions.

In a CISC microcontroller both data and instructions are 8-bits wide. CISC microcontrollers usually have over 200 instructions. Data and code are on the same bus and cannot be fetched simultaneously.

1.5   Exercises

1. What is a microcontroller? What is a microprocessor? Explain the main differences between a microprocessor and a microcontroller.
2. Give some example applications of microcontrollers around you.
3. Where would you use an EPROM memory?
4. Where would you use a RAM memory?
5. Explain what type of memories are usually used in microcontrollers.
6. What is an I/O port?
7. What is an analogue-to-digital converter? Give an example use for this converter.

8. Explain why a watchdog timer could be useful in a real-time system.
9. What is serial I/O? Where would you use serial communication?
10. Why is the current sinking/sourcing important in the specification of an output port pin?
11. What is an interrupt? Explain what happens when an interrupt is recognised by a microcontroller.
12. Why is brown-out detection important in real-time systems?
13. Explain the differences between a RISC-based microcontroller and a CISC-based microcontroller. What type of microcontroller is PIC?

2
The PIC microcontroller family

The PIC microcontroller family of microcontrollers is manufactured by *Microchip Technology Inc.* Currently they are one of the most popular microcontrollers used in many commercial and industrial applications. Over 120 million devices are sold each year.

The PIC microcontroller architecture is based on a modified Harvard RISC (Reduced Instruction Set Computer) instruction set with dual-bus architecture, providing fast and flexible design with an easy migration path from only 6 pins to 80 pins, and from 384 bytes to 128 kbytes of program memory.

PIC microcontrollers are available with many different specifications depending on:

- Memory Type
  - Flash
  - OTP (One-time-programmable)
  - ROM (Read-only-memory)
  - ROMless
- Input–Output (I/O) Pin Count
  - 4–18 pins
  - 20–28 pins
  - 32–44 pins
  - 45 and above pins
- Memory Size
  - 0.5–1 K
  - 2–4 K
  - 8–16 K
  - 24–32 K
  - 48–64 K
  - 96–128 K
- Special Features
  - CAN
  - USB
  - LCD
  - Motor Control
  - Radio Frequency

Although there are many models of PIC microcontrollers, the nice thing is that they are upward compatible with each other and a program developed for one model can very easily, and in many cases with no modifications, be run on other models of the family. The basic assembler instruction set of PIC microcontrollers consists of only 33 instructions and most of the family members (except the newly developed devices) use the same instruction set. This is why a program developed for one model can run on another model with similar architecture without any changes.

All PIC microcontrollers offer the following features:

- RISC instruction set with only a handful of instructions to learn
- Digital I/O ports
- On-chip timer with 8-bit prescaler
- Power-on reset
- Watchdog timer
- Power saving SLEEP mode
- High source and sink current
- Direct, indirect, and relative addressing modes
- External clock interface
- RAM data memory
- EPROM or Flash program memory

Some devices offer the following additional features:

- Analogue input channels
- Analogue comparators
- Additional timer circuits
- EEPROM data memory
- External and internal interrupts
- Internal oscillator
- Pulse-width modulated (PWM) output
- USART serial interface

Some even more complex devices in the family offer the following additional features:

- CAN bus interface
- I^{2}C bus interface
- SPI bus interface
- Direct LCD interface
- USB interface
- Motor control

Although there are several hundred models of PIC microcontrollers, choosing a microcontroller for an application is not a difficult task and requires taking into account these factors:

- Number of I/O pins required
- Required peripherals (e.g. USART, USB)

- The minimum size of program memory
- The minimum size of RAM
- Whether or not EEPROM non-volatile data memory is required
- Speed
- Physical size
- Cost.

The important point to remember is that there could be many models which satisfy all of the above requirements. You should always try to find the model which satisfies your minimum requirements and the one which does not offer more than you may need. For example, if you require a microcontroller with only 8 I/O pins and if there are two identical microcontrollers, one with 8 and the other one with 16 I/O pins, you should select the one with 8 I/O pins.

Although there are several hundred models of PIC microcontrollers, the family can be broken down into three main groups, which are:

- 12-bit instruction word (e.g. 12C5XX, 16C5X)
- 14-bit instruction word (e.g. 16F8X, 16F87X)
- 16-bit instruction word (e.g. 17C7XX, 18C2XX).

All three groups share the same RISC architecture and the same instruction set, with a few additional instructions available for the 14-bit, and many more instructions available for the 16-bit models. Instructions occupy only one word in memory, thus increasing the code efficiency and reducing the required program memory. Instructions and data are transferred on separate buses, thus the overall system performance is increased.

The features of some microcontrollers in each group are given in the following sections.

2.1   12-bit instruction word

Table 2.1 lists some of the devices in this group. Because of the simplicity of their architecture these devices are not supported by the PicBasic compiler. PicBasic Pro compiler provides a limited support for these devices. But, as the prices of 14-bit devices have declined, there really is no need anymore to use a 12-bit device, except for their smaller physical sizes.

PIC12C508: This is a low-cost, 8-pin device with 512×12 EPROM program memory, and 25 bytes of RAM data memory. The device can operate at up to 4 MHz clock input and the instruction set consists of only 33 instructions. The device features 6 I/O ports, 8-bit timer, power-on reset, watchdog timer, and internal 4 MHz oscillator capability. One of the major disadvantages of this microcontroller is that the program memory is EPROM-based and it cannot be erased or programmed using the standard programming devices. The program memory has to be erased using an EPROM eraser device (an ultraviolet light source).

The "F" version of this family (e.g. PIC12F508) is based on flash program memory which can be erased and re-programmed using the standard PIC programmer devices. Similarly, the "CE" version of the family (e.g. PIC12CE518) offers an additional 16-byte non-volatile EEPROM data memory.

Table 2.1   Some 12-bit PIC microcontrollers

Microcontroller	Program Memory	Data RAM	Max speed (MHz)	I/O Ports	A/D Converter
12C508	512 × 12	25	4	6	–
16C54	384 × 12	25	20	12	–
16C57	2048 × 12	72	20	20	–
16C505	1024 × 12	41	4	12	–
16C58A	2048 × 12	73	20	12	–

Figure 2.1 shows the pin configuration of the PIC12F508 microcontroller.

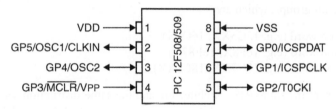

Figure 2.1   PIC12F508 microcontroller

PIC16C5X: This is one of the earliest PIC microcontrollers. The device is 18-pin with a 384 × 12 EPROM program memory, 25 bytes of RAM data memory, 12 I/O ports, a timer, and a watchdog. Some other members in the family, e.g. PIC16C56 have the same architecture but more program memory (1024 × 12). PIC16C58A has more program memory (2048 × 12) and also more data memory (73 bytes of RAM). Figure 2.2 shows the pin configuration of the PIC16C56 microcontroller.

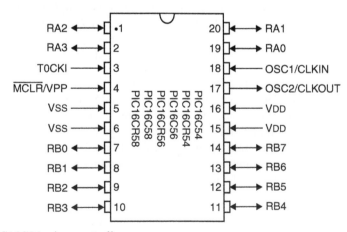

Figure 2.2   PIC16C56 microcontroller

2.2   14-bit instruction word

This is a big family including many models of PIC microcontrollers. These devices are supported by both the PicBasic and PicBasic Pro compilers. Most of the devices in this family can operate at up to 20 MHz clock rate. The instruction set consists of 35 instructions. These devices offer advanced features such as internal and external interrupt sources. Table 2.2 lists some of the microcontrollers in this group.

Table 2.2   Some 14-bit microcontrollers

Microcontroller	Program Memory	Data RAM	Max speed (MHz)	I/O Ports	A/D Converter
16C554	512 × 14	80	20	13	–
16C64	2048 × 14	128	20	33	–
16F84	1024 × 14	36	10	13	–
16F627	1024 × 14	224	20	16	–
16F628	2048 × 14	224	20	16	–
16F676	1024 × 14	64	20	12	8
16F73	4096 × 14	192	20	22	5
16F876	8192 × 14	368	20	22	5
16F877	8192 × 14	368	20	33	8

PIC16C554: This microcontroller has similar architecture to the PIC16C54 but the instructions are 14 bits wide. The program memory is 512×14 and the data memory is 80 bytes of RAM. There are 13 I/O pins where each pin can source or sink 25 mA current. Additionally, the device contains a timer and a watchdog.

PIC16F84: This has been one of the most popular PIC microcontrollers for a very long time. This is an 18-pin device and it offers 1024×14 flash program memory, 36 bytes of data RAM, 64 bytes of non-volatile EEPROM data memory, 13 I/O pins, a timer, a watchdog, and internal and external interrupt sources. The timer is 8-bits wide but can be programmed to generate internal interrupts for timing purposes. PIC16F84 can be operated from a crystal or a resonator for accurate timing. A resistor-capacitor can also be used as a timing device for applications where accurate timing is not required.

We will be using the PIC16F84 in some of the projects in this book. Figure 2.3 shows the pin configuration of this microcontroller. The pin descriptions are given in Table 2.3.

PIC16F877: This microcontroller is a 40-pin device and is one of the popular microcontrollers used in complex applications. The device offers 8192×14 flash program memory, 368 bytes of RAM, 256 bytes of non-volatile EEPROM memory, 33 I/O pins, 8 multiplexed A/D converters with 10-bits resolution, PWM generator, 3 timers, analogue capture and comparator circuit, USART, and internal and external interrupt facilities.

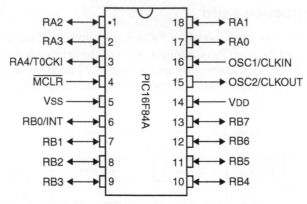

Figure 2.3   PIC16F84 microcontroller pin configuration

Table 2.3   PIC16F84 microcontroller pin descriptions

Pin	Description	Pin	Description
1	RA2 – PORTA bit 2	10	RB4 – PORTB bit 4
2	RA3 – PORTA bit 3	11	RB5 – PORTB bit 5
3	RA4/T0CK1 – PORTA bit 4/Counter clk	12	RB6 – PORTB bit 6
4	MCLR – Master clear	13	RB7 – PORTB bit 7
5	Vss – Gnd	14	Vdd – +V supply
6	RB0/INT – PORTB bit 0	15	OSC2
7	RB1 – PORTB bit 1	16	OSC1
8	RB2 – PORTB bit 2	17	RA0 – PORTA bit 0
9	RB3 – PORTB bit 3	18	RA1 – PORTA bit 1

We will be using the PIC16F877 in some of the projects in this book. Figure 2.4 shows the pin configuration of this microcontroller.

PIC16F627: This is an 18-pin microcontroller with 1024 × 14 flash program memory. The device offers 224 bytes of RAM, 128 bytes of non-volatile EEPROM memory, 16 I/O pins, two 8-bit timers, one 16-bit timer, a watchdog, and comparator circuits. This microcontroller is similar to PIC16F84, but offers more I/O pins, more program memory, and a lot more RAM. In addition, PIC16F627 is more suited to applications which require more than one timer.

We will be using the PIC16F627 in some of the projects in this book. Figure 2.5 shows the pin configuration of this microcontroller.

PIC16F676: This is a 14-pin microcontroller which is becoming very popular. The device offers 1024 × 14 flash program memory, 64 bytes of RAM, 12 I/O pins, 128 bytes of EEPROM, 8

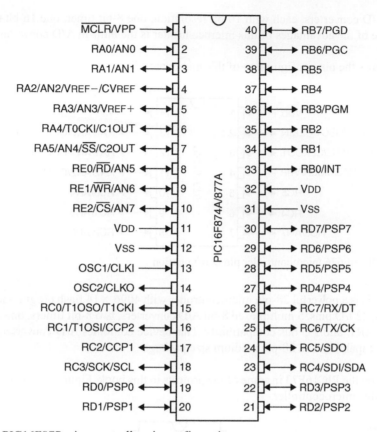

Figure 2.4   PIC16F877 microcontroller pin configuration

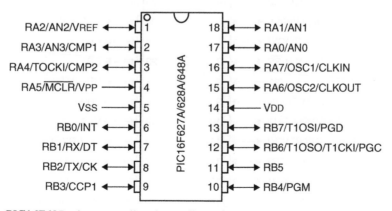

Figure 2.5   PIC16F627 microcontroller pin configuration

multiplexed A/D converters, each with 10-bit resolution, one 8-bit timer, one 16-bit timer, and a watchdog. One of the advantages of this microcontroller is the built-in A/D converter.

Figure 2.6 Shows the pin configuration of this microcontroller.

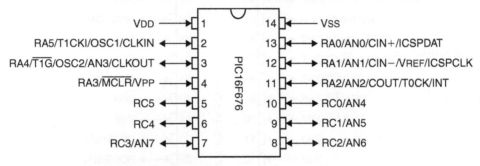

Figure 2.6   PIC16F676 microcontroller pin configuration

PIC16F73: This is a powerful 28-pin microcontroller with 4096 ×14 flash program memory, 192 bytes of RAM, 22 I/O pins, 5 multiplexed 8-bit A/D converters, two 8-bit timers, one 16-bit timer, watchdog, USART, and I²C bus compatibility. This device combines A/D converter, digital I/O, and serial I/O capability in a 28-pin medium size package.

We will be using the PIC16F73 in some of the projects in this book. Figure 2.7 shows the pin configuration of this microcontroller.

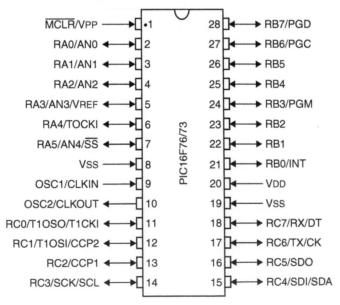

Figure 2.7   PIC16F73 microcontroller pin configuration

2.3 16-bit instruction word

16-bit microcontrollers are at the high-end of the PIC microcontroller family. These microcontrollers cannot be used with the PicBasic compiler, but the PicBasic Pro can be used to program them. Most of the devices in this group can operate at up to 40 MHz, have 33 I/O pins, and 3 timers. They have 23 instructions in addition to the 35 instructions found on the 14-bit microcontrollers. Table 2.4 lists some of the devices in this family. We will not be using any of the 16-bit microcontrollers in the projects in this book, and I won't spend more time to describe the features of this group. Interested readers can look at the Microchip web site at www.microchip.com.

Table 2.4   Some 16-bit microcontrollers

Microcontroller	Program Memory	Data RAM	Max speed (MHz)	I/O Ports	A/D Converter
17C43	4096 × 16	454	33	33	–
17C752	8192 × 16	678	33	50	12
18C242	8192 × 16	512	40	23	5
18C252	16384 × 16	1536	40	23	5
18C452	16384 × 16	1536	40	34	8

All memory for the PIC microcontroller family is internal and it is usually not very easy to extend this memory externally. No special hardware or software features are provided for extending either the program memory or the data memory. The program memory is usually sufficient for small to medium size projects. But the data memory is generally small and may not be enough for medium to large projects unless a bigger and more expensive member of the family is chosen. For some large projects even this may not be enough and the designer may have to sacrifice the I/O ports to interface an external data memory, or to choose a microcontroller from a different manufacturer.

2.4   Inside a PIC microcontroller

Although there are many models of microcontrollers in the PIC family, they all share some common features, such as program memory, data memory, I/O ports, and timers. Some devices have additional features such as A/D converters, USART and so on. Because of these common features, we can look at these attributes and cover the operation of most devices in the family.

2.4.1   Program memory (Flash)

The program memory is where your PicBasic or PicBasic Pro program resides. In early microprocessors and microcontrollers the program memory was EPROM which meant that it had to be erased using UV light before it could be re-programmed. Most PIC microcontrollers nowadays are based on the flash technology where the memory chip can be erased or re-programmed using a programmer device. Most PIC microcontrollers can also be programmed without removing them from their circuits. This process (called in-circuit serial programming, or ISP) speeds up the development cycle and lowers the development costs. Although the program memory is mainly used to store a program, there is no reason why it cannot be used to store constant data used in programs.

PIC microcontrollers can have program memories from 0.5 to over 16 K. A PicBasic program can have several pages of code and still fit inside a 1 K of program memory. The width of a 14-bit program memory is actually 14 bits wide. It is interesting to note that PICs are known as 8-bit microcontrollers. This is actually true as far as the width of the data memory is concerned, which is 8-bits wide. Microchip calls the 14-bits a *word*, even though a *word* is actually 16-bits wide.

Although the size of the program memory can be larger than 2 K, PicBasic compiler can only work with the first 2 K which can be a limiting factor in large projects. PicBasic Pro compiler can use all the available program memory space.

When power is applied to the microcontroller or when the MCLR input is lowered to logic 0, execution start from the Reset Vector, which is the first word of the program memory. Thus, the first instruction executed after a reset is the one located at address 0 of the program memory. When the program is written in assembler language the programmer has to use special instructions (called ORG) so that the first executable instruction is loaded into address 0 of the program memory. High-level languages such as PicBasic or PicBasic Pro compile your program such that the first executable statement in your program is loaded into the first location of the program memory.

2.4.2   Data memory (RAM)

The data memory is used to store all of your program variables. This is a RAM which means that all the data is lost when power is removed. The width of the data memory is 8-bits wide and this is why the PIC microcontrollers are called 8-bit microcontrollers.

The data memory in a PIC microcontroller consists of banks where some models may have only 2 banks, some models 4 banks, and so on. A required bank of the data memory can be selected under program control.

2.4.3   Register file map and special function registers

Register File Map (RFM) is a layout of all the registers available in a microcontroller and this is extremely useful when programming the device, especially when using an assembler language. The RFM is divided into two parts: the *Special Function Registers* (SFR), and the *General Purpose Registers* (GPR). For example, on a PIC16F84 microcontroller there are 68 GPR registers and these are used to store temporary data. We shall see later on when programming in PicBasic or PicBasic Pro that these registers are used to store the variables declared in a program.

SFR is a collection of registers used by the microcontroller to control the internal operations of the device. Depending upon the complexity of the devices the number of registers in the SFR varies. It is important that the programmer understands the functions of the SFR registers fully since they are used both in assembly language and in high-level languages.

Depending on the model of PIC microcontroller used there could be other registers. You need not know the operation of some of the registers since PicBasic and PicBasic Pro compiler loads these registers automatically. For example, writing and reading from the EEPROM are controlled by

SFR registers EECON1, EECON2, EEADR, and EEDATA. But fortunately, PicBasic and PicBasic Pro compilers provide simple high-level instructions for writing to and reading from the EEPROM and thus you do not need to know how to load these registers.

Some of the important SFR registers that you may need to configure while programming using a high-level language are

- OPTION register
- I/O registers
- Timer registers
- INTCON register
- A/D converter registers

The functions and the bit definitions of these registers are described in detail in the following sections.

OPTION register

This register is used to setup various internal features of the microcontroller and is named as OPTION_REG. This is a readable and writable register which contains various control bits to configure the on-chip timer and the watchdog timer. This register is at address 81 (hexadecimal) of the microcontroller and its bit definitions are given in Figure 2.8. The OPTION REG register is also used to control the external interrupt pin RB0. This pin can be setup to generate an interrupt, for example, when it changes from logic 0 to logic 1. The microcontroller then suspends the main program execution and jumps to the interrupt service routine (ISR) to service the interrupt. Upon return from the interrupt, normal processing resumes.

For example, to configure the INT pin so that external interrupts are accepted on the rising edge of the INT pin, the following bit pattern should be loaded into the OPTION_REG:

> X1XXXXXX

Where X is a don't care bit and can be a 0 or a 1. We shall see in the projects section on how to configure various bits of this register.

I/O registers

These registers are used for the I/O control. Every I/O port in the PIC microcontroller has two registers: *port data register* and *port direction control register*.

Port data register has the same name as the port it controls. For example, PIC16F84 microcontroller has two port data registers PORTA and PORTB. A PIC16F877 microcontroller has 5 port data registers PORTA, PORTB, PORTC, PORTD, and PORTE. An 8-bit data can be sent to any port, or an 8-bit data can be read from the ports. It is also possible to read or write to individual port pins. For example, any bit of a given port can be set or cleared, or data can be read from one or more port pins at the same time.

7	6	5	4	3	2	1	0
RBPU	INTEDG	T0CS	T0SE	PSA	PS2	PS1	PS0

Bit 7: PORTB Pull-up Enable
    1: PORTB pull-ups disabled
    0: PORTB pull-ups enabled

Bit 6: INT Interrupt Edge Detect
    1: Interrupt on rising edge of INT input
    0: Interrupt on falling edge of INT input

Bit 5: TMR0 Clock Source
    1: T0CK1 pulse
    0: Internal oscillator

Bit 4: TMR0 Source Edge Select
    1: Increment on HIGH to LOW of T0CK1
    0: Increment on LOW to HIGH of T0CK1

Bit 3: Prescaler Assignment
    1: Prescaler assigned to Watchdog Timer
    0: Prescaler assigned to TMR0

Bit 2-0: Prescaler Rate
    000   1:2
    001   1:4
    010   1:8
    011   1:16
    100   1:32
    101   1:64
    110   1:128
    111   1:256

Figure 2.8   OPTION_REG bit definitions

Ports in a PIC microcontroller are bi-directional. Thus, each pin of a port can be used as an input or an output pin. Port direction control register configures the port pins as either inputs or outputs. This register is called the TRIS register and every port has a TRIS register named after its port name. For example, TRISA is the direction control register for PORTA. Similarly, TRISB is the direction control register for PORTB and so on.

Setting a bit in the TRIS register makes the corresponding port register pin an input. Clearing a bit in the TRIS register makes the corresponding port pin an output. For example, to make bits 0 and 1 of PORTB input and the other bits output, we have to load the TRISB register with the bit pattern.

00000011

Figure 2.9 shows the TRISB register and the direction of PORTB pins.

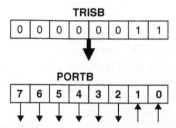

Figure 2.9 TRISB and PORTB direction

Port data register and port direction control registers can be accessed directly using the PicBasic Pro compiler. For example, as we shall see in a later chapter, TRISB register can be set to 3 and data can be read from PORTB into a variable named CNT by the PicBasic Pro instructions.

TRISB = 3
CNT = PORTB

The PicBasic compiler has no direct register control instructions and as we shall see in a later chapter, we have to use the PEEK and POKE instructions. PEEK is used to read data from a register and POKE is used to send data to a register.

When we use the PEEK and POKE instructions we have to specify the register address of the register we wish to access. The register addresses of port registers are (the "$" character specifies that the number is in hexadecimal format)

Ports	Address (Hexadecimal)
PORTA	$05
PORTB	$06
PORTC	$07
PORTD	$08
PORTE	$09
TRISA	$85
TRISB	$86
TRISC	$87
TRISD	$88
TRISE	$89

Thus, for the above example, the required PicBasic instructions will be

    POKE $86, 3
    PEEK $06, CNT

We shall see in the next chapter how to use symbols in PicBasic language to make our programs clearer and easier to maintain.

Timer registers

Depending on the model used, some PIC microcontrollers have only one timer, and some may have up to 3 timers. In this section we shall look at the PIC16F84 microcontroller which has only one timer. The extension to several timers is similar and we shall see in the projects section how to use more than one timer.

The timer in the PIC16F84 microcontroller is an 8-bit register (called TMR0) which can be used as a timer or a counter. When used as a counter, the register increments each time a clock pulse is applied to pin T0CK1 of the microcontroller. When used as a timer, the register increments at a rate determined by the system clock frequency and a prescaler selected by register OPTION_REG. Prescaler rates vary from 1:2 to 1:256. For example, when using a 4 MHz clock, the basic instruction cycle is 1 μs (the clock is internally divided by four). If we select a prescaler rate of 1:16, the counter will be incremented at every 16 μs.

The TMR0 register has address 01 in the RAM which can be loaded using the POKE instruction in PicBasic, or by accessing the TMR0 register directly in PicBasic Pro.

A timer interrupt is generated when the timer overflows from 255 to 0. This interrupt can be enabled or disabled by our program. Thus, for example, if we require to generate interrupts at 200 μs intervals using a 4 MHz clock, we can select a prescaler value of 1:4 and enable timer interrupts. The timer clock rate is then 4 μs. For a time-out of 200 μs, we have to send 50 clocks to the timer. Thus, the TMR0 register should be loaded with $256 - 50 = 206$, i.e. a count of 50 before an overflow occurs.

The watchdog timer's oscillator is independent from the CPU clock and the time-out is 18 ms. To prevent a time-out condition the watchdog must be reset periodically via software. If the watchdog timer is not reset before it times out, the microprocessor will be forced to jump to the reset address. The prescaler can be used to extend the time-out period and valid rates are 1, 2, 4, 8, 16, 32, 64, and 128. For example, when set to 128, the time out period is about 2 s ($18 \times 128 = 2304$ ms). The watchdog timer can be disabled during programming of the device if it is not used.

Since the timer is very important part of the PIC microcontrollers more detailed information is given on its operation below.

TMR0 and watchdog

TMR0 and a watchdog are found nearly in all PIC microcontrollers. Figure 2.10 shows the functional diagram of TMR0 and the watchdog circuit. The operation of the watchdog circuit is as described earlier and only the TMR0 circuit is described in this section.

The source of input for TMR0 is selected by bit T0CS of OPTION_REG and it can be either from the microcontroller oscillator f_{osc} divided by 4, or it can be an external clock applied to the RA4/T0CK1 input. Here, we will only look at using the internal oscillator. If a 4 MHz crystal is used the internal oscillator frequency is $f_{osc}/4 = 1$ MHz which corresponds to a period of $T = 1/f = 10^{-6}$, or 1 μs. TMR0 is then selected as the source for the prescaler by clearing PSA bit of OPTION_REG. The required prescaler value is selected by bits PS0 to PS2 as shown in Figure 2.8. Bit PSA should then be cleared to 0 to select the prescaler for the timer. All the bits are configured now and TMR0 register increments each time a pulse is applied by the internal oscillator. TMR0 register is 8-bits wide and it counts up to 255, then creates an overflow condition, and continues counting from 0. When TMR0 changes from 255 to 0 it generates a timer interrupt if timer interrupts and global interrupts are enabled (see INTCON register. Interrupt will be generated if GIE and TMR0 bits of INTCON are both set to 1). See the Section 2.4.6 on Interrupts for more information.

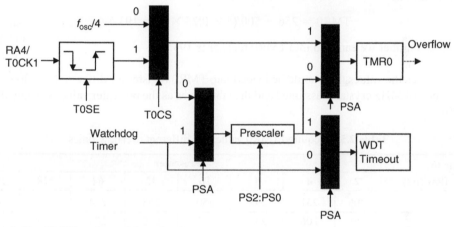

Figure 2.10   TMR0 and watchdog circuit

By loading a value into the TMR0 register we can control the count until an overflow occurs. The formula given below can be used to calculate the time it will take for the timer to overflow (or to generate an interrupt) given the oscillator period, value loaded into the timer and the prescaler value.

$$\text{Overflow time} = 4 \times T_{osc} \times \text{Prescaler} \times (256 - \text{TMR0}) \qquad (2.1)$$

where

Overflow time	is in μs,
T_{osc}	is the oscillator period in μs,
Prescaler	is the prescaler value chosen using OPTION_REG
TMR0	is the value loaded into TMR0 register.

For example, assume that we are using a 4 MHz crystal, and the prescaler chosen as 1:8 by setting bits PS2:PS0 to "010". Also assume that the value loaded into the timer register TMR0 is decimal 100. The overflow time is then given by

$$4\,\text{MHz clock has a period}, T = 1/f = 0.25\,\mu s$$

Using the above formula,

$$\text{Overflow time} = 4 \times 0.25 \times 8 \times (256 - 100) = 1248\,\mu s.$$

Thus, the timer will overflow after 1.248 ms and a timer interrupt will be generated if the timer interrupt and global interrupts are enabled.

What we normally need is to know what value to load into the TMR0 register for a required Overflow time. This can be calculated by modifying Eq. (2.1) as

$$\text{TMR0} = 256 - (\text{Overflow time})/(4 \times T_{\text{osc}} \times \text{Prescaler}) \tag{2.2}$$

For example, suppose that we want an interrupt to be generated after 500 μs and the clock and the prescaler values are as before. The value to be loaded into the TMR0 register can be calculated using Eq. (2.2) as

$$\text{TMR0} = 256 - 500/(4 \times 0.25 \times 8) = 193.5$$

The nearest number we can load into TMR0 register is 193.

Table 2.5 gives the values that should be loaded into TMR0 register for different Overflow times. In this table a 4 MHz crystal is assumed and the table gives as the prescaler value is changed from 2 to 256.

Table 2.5   Required TMR0 values for different overflow times

Time to overflow (μs)	Prescaler							
	2	4	8	16	32	64	128	256
100	206	231	243	250	253	254	–	–
200	156	206	231	243	250	253	254	–
300	106	181	218	237	246	251	253	255
400	56	156	206	231	243	250	253	254
500	6	131	193	224	240	248	252	254
600	–	106	181	218	237	16	251	253
700	–	81	168	212	234	245	250	253
800	–	56	156	206	231	243	250	253
1,000	–	6	131	193	225	240	248	252
5,000	–	–	–	–	100	178	77	236
10,000	–	–	–	–	–	100	178	217
20,000	–	–	–	–	–	–	100	178
30,000	–	–	–	–	–	–	–	139
40,000	–	–	–	–	–	–	–	100
50,000	–	–	–	–	–	–	–	60
60,000	–	–	–	–	–	–	–	21

TMR1

Although TMR0 is the basic timer found nearly in all PIC microcontrollers, some devices have several timers, e.g. TMR0, TMR1, and TMR2. Additional timers give added functionality to a microcontroller. In this section the operation of TMR1 will be described in detail.

TMR1 is a 16-bit timer, consisting of two 8-bit registers TMR1H and TMR1L. As shown in Figure 2.11, a prescaler is used with TMR1 and the available prescaler values are only 1, 2, 4, and 8.

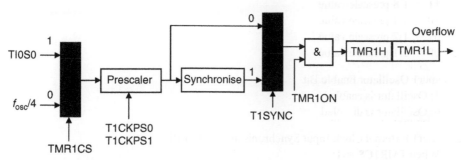

Figure 2.11   TMR1 structure

Register T1CON controls the operation of TMR1. The bit definition of this register is shown in Figure 2.12. TMR1 can operate either as a timer or as a counter, selected by bit TMR1CS of T1CON. When operated in timer mode, TMR1 increments every oscillator frequency $f_{osc}/4$. TMR1 can be enabled or disabled by setting or clearing control bit TMR1ON. TMR1 can count from 0 to 65,535 and it generates an overflow when changing from 65,535 to 0. A timer interrupt is generated if the TMR1 interrupt enable bit TMR1IE is enabled and also the global interrupts are enabled by register INTCON.

When TMR1 is operated in counter mode, it increments on every rising edge (from logic 0 to logic 1) of the clock input.

TMR2

TMR2 is an 8-bit timer with a prescaler and a postscaler and it has an 8-bit period register PR2. This timer is controlled by register T2CON whose bit definitions are given in Figure 2.13. The prescaler options are 1, 4, and 16 only and are selected by T2CKPS1 and T2CKPS0 bits of T2CON. TMR2 increments from 0, until it matches PR2, and then resets to 0 on the next cycle. Then the cycle is repeated. TMR2 can be shut off by clearing TMR2ON of T2CON register to minimise power consumption.

INTCON register

This is the interrupt control register. This register is at address 0 and 8B (hexadecimal) of the microcontroller RAM and the bit definitions are given in Figure 2.14. For example, to enable

7	6	5	4	3	2	1	0
–	–	TICKPS1	TICKPS0	TIOSCEN	TISYNC	TMR1CS	TMR1ON

Bit 7: Unused

Bit 6: Unused

Bit 5-4: Timer1 Input Clock Prescale Select Bits
      11   1:8 prescale value
      10   1:4 prescale value
      01   1:2 prescale value
      00   1:1 prescale value

Bit 3: Timer1 Oscillator Enable Bit
      1: Oscillator is enabled
      0: Oscillator is disabled

Bit 2: Timer1 External Clock Input Synchronisation Select Bit
      When TMR1CS = 1:
      1: Do not synchronise external clock input
      0: Synchronise external clock input
      When TMR1CS = 0:
      This bit is ignored. Timer1 uses internal clock

Bit 1: Timer1 Clock Source Select Bit
      1: External clock from pin TIOSO (on rising edge)
      0: Internal clock ($f_{osc}/4$)

Bit 0: Timer1 On Bit
      1: Enable Timer1
      0: Stops Timer1

Figure 2.12   T1CON bit definitions

interrupts so that external interrupts from pin INT (RB0) can be accepted on a PIC16F84, the following bit pattern should be loaded into register INTCON:

      1XX1XXXX

Similarly, to enable timer interrupts, bit 5 of INTCON must be set to 1.

A/D converter registers

The A/D converter is used to interface analogue signals to the microcontroller. The A/D converts analogue signals (e.g. voltage) into digital form so that they can be connected to a computer. A/D converter registers are used to control the A/D converter ports. On most PIC microcontrollers equipped with A/D, PORTA pins are used for analogue input and these port pins are shared between digital and analogue functions.

7	6	5	4	3	2	1	0
–	TOUTPS3	TOUTPS2	TOUTPS1	TOUTPS0	TMR2ON	T2CKPS1	T2CKPS0

Bit 7: Unused

Bit 6-3: Timer2 Output Postscale Select Bits
        0000   1:1 Postscale
        0001   1:2 Postscale
        0010   1:3 Postscale
        ....
        ....
        1111   1:16 Postscale

Bit 2: Timer2 On Bit
        1: Timer2 is On
        0: Timer2 is Off

Bit 1-0: Timer2 Clock Prescale Select Bits
        00   Prescaler is 1
        01   Prescaler is 4
        10   Prescaler is 16
        11   Prescaler is 16

Figure 2.13   T2CON bit definitions

PIC16F876 includes 5 A/D converters. Similarly, PIC16F877 includes 8 A/D converters. There is actually only one A/D converter as shown in Figure 2.15 and the inputs are multiplexed and they share the same converter. The width of the A/D converter can be 8-bits or 10-bits. Both PIC16F876 and PIC16F877 have 10-bit converters. PIC16F73 has 8-bit converters. An A/D converter requires a reference voltage to operate. This reference voltage is chosen by programming the A/D converter registers and is typically +5 V. Thus, if we are using a 10-bit converter (1024 quantisation levels) the resolution of our converter will be 5/1024 = 0.00488 V, or 4.88 mV, i.e. we can measure analogue voltages with a resolution of 4.88 mV. For example, if the measured analogue input voltage is 4.88 mV we get the 10-bit digital number "0000000001", if the analogue input voltage is 2 × 4.88 = 9.76 mV, the 10-bit converted number will be "0000000010", if the analogue input voltage is 3 × 4.88 = 14.64 mV, the converted number will be "0000000011", and so on.

In a similar way, if the reference voltage is +5 V and we are using an 8-bit converter (256 quantisation levels), the resolution of the converter will be 5/256 = 19.53 mV. For example, if the measured input voltage is 19.53 mV we get the 8-bit number "00000001", if the analogue input voltage is 2 × 19.53 = 39.06 mV we get the 8-bit number "00000010", and so on.

The A/D converter is controlled by registers ADCON0 and ADCON1. The bit pattern of ADCON0 is shown in Figure 2.16. ADCON0 is split into four parts, the first part consists of the highest two

7	6	5	4	3	2	1	0
GIE	EEIE	T0IE	INTE	RBIE	T0IF	INTF	RBIF

Bit 7: Global Interrupt Enable
> 1: Enable all un-masked interrupts
> 0: Disable all interrupts

Bit 6: EE Write Complete Interrupt
> 1: Enable EE write complete interrupt
> 0: Disable EE write complete interrupt

Bit 5: TMR0 Overflow Interrupt
> 1: Enable TMR0 interrupt
> 0: Disable TMR0 interrupt

Bit 4: INT External Interrupt
> 1: Enable INT External Interrupt
> 0: Disable INT External Interrupt

Bit 3: RB Port Change Interrupt
> 1: Enable RB port change interrupt
> 0: Disable RB port change interrupt

Bit 2: TMR0 Overflow Interrupt Flag
> 1: TMR0 has overflowed
> 0: TMR0 did not overflow

Bit 1: INT Interrupt Flag
> 1: INT interrupt occurred
> 0: INT interrupt did not occur

Bit 0: RB Port Change Interrupt Flag
> 1: One or more RB4-RB7 pins changed state
> 0: None of RB4-RB7 changed state

Figure 2.14   INTCON register bit definitions

bits ADCS1 and ADCS0 and they are used to select the conversion clock. The internal RC oscillator or the external clock can be selected as the conversion clock as in the following table:

00	External clock/2
01	External clock/8
10	External clock/32
11	Internal RC clock

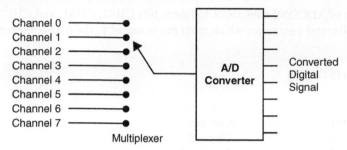

Figure 2.15　Multiplexed A/D structure

7	6	5	4	3	2	1	0
ADCS1	ADCS0	CHS2	CHS1	CHS0	GO/DONE	–	ADON

Bit 7-6: A/D Converter Clock Select

      00     $f_{osc}/2$
      01     $f_{osc}/8$
      10     $f_{osc}/32$
      11   Internal RC oscillator

Bit 5-3: A/D Channel Select

      000   Channel 0
      001   Channel 1
      010   Channel 2
      011   Channel 3
      100   Channel 4
      101   Channel 5
      110   Channel 6
      111   Channel 7

Bit 2: GO/DONE Bit

      1: Start conversion
      0: A/D conversion is complete

Bit 1: Not used

Bit 0: ADON Bit

      1: Turn ON A/D circuit
      0: Turn OFF A/D circuit

Figure 2.16　ADCON0 bit definitions

The second part of ADCON0 consists of the three bits CHS2, CHS1, and CHS0. These are the channel select bits, and they select which input pin is routed to the A/D converter. The selection is as follows:

CHS2:CHS1:CHS0

000	Channel 0
001	Channel 1
010	Channel 2
011	Channel 3
100	Channel 4
101	Channel 5
110	Channel 6
111	Channel 7

The third part of ADCON0 is the single GO/DONE bit. This bit has two functions: first, by setting the bit it starts the A/D conversion. Second, the bit is cleared when the conversion is complete and this bit can be checked to see whether or not the conversion is complete.

The fourth part of ADCON0 is also a single bit ADON which is set to turn on the A/D converter circuitry.

ADRESH and ADRESL are the A/D converter result registers. ADRESL is the low byte and ADRESH is the upper 2 bits (if a 10-bit converter is used). We shall see how to configure the result of the conversion later.

ADCON1 is the second A/D control register. This register controls the format of converted data and mode of the PORTA inputs. The bit format of this register is shown in Figure 2.17. Bit 7 is called ADFM and when this bit is 0 the result of the A/D conversion is left justified, when it is 1, the result of the A/D conversion is right justified. If we have an 8-bit converter we can clear ADFM and just read ADRESH to get the 8-bit converted data. If we have a 10-bit converter we can set ADFM to 1 and the 8 bits of the result will be in ADRESL, 2 bits of the result will be in the lower bit positions of ADRESH. The remaining 6 positions of ADRESH (bit 2 to bit 7) will be cleared to zero.

Bits PCFG0-3 control the mode of PORTA pins. As seen in Figure 2.17, a PORTA pin can be programmed to be a digital pin or an analogue pin. For example, if we set PCFG0-3 to "0110" then all PORTA pins will be digital I/O pins. PCFG0-3 bits can also be used to define the reference voltage for the A/D converter. As we shall see in the projects section of the book, the reference voltage Vref+ is usually set to be equal to the supply voltage (Vdd), and Vref− is set to be equal to Vss. This makes the A/D reference voltage to be +5 V.

2.4.4   Oscillator circuits

An Oscillator circuit is used to provide a microcontroller with a clock. A clock is needed so that the microcontroller can execute a program.

7	6	5	4	3	2	1	0
ADFM	–	–	–	PCFG3	PCFG2	PCFG1	PCFG0

Bit 7: A/D Converter Result Format Select
    1: A/D converter output is right justified
    0: A/D converter output is left justified

Bit 6: Not used

Bit 5: Not used

Bit 4: Not used

Bit 3-0: Port Assignment and Reference Voltage Selection
    (see Table below)

PCFG3-PCFG0	AN7	AN6	AN5	AN4	AN3	AN2	AN1	AN0	Vref+	Vref−
0000	A	A	A	A	A	A	A	A	Vdd	Vss
0001	A	A	A	A	Vref+	A	A	A	RA3	Vss
0010	D	D	D	A	A	A	A	A	Vdd	Vss
0011	D	D	D	A	Vref+	A	A	A	RA3	Vss
0100	D	D	D	D	A	D	A	A	Vdd	Vss
0101	D	D	D	D	Vref+	D	A	A	RA3	Vss
0110	D	D	D	D	D	D	D	D	Vdd	Vss
0111	D	D	D	D	D	D	D	D	Vdd	Vss
1000	A	A	A	A	Vref+	Vref−	A	A	RA3	RA2
1001	D	D	A	A	A	A	A	A	Vdd	Vss
1010	D	D	A	A	Vref+	A	A	A	RA3	Vss
1011	D	D	A	A	Vref+	Vref−	A	A	RA3	RA2
1100	D	D	D	A	Vref+	Vref−	A	A	RA3	RA2
1101	D	D	D	D	Vref+	Vref−	A	A	RA3	RA2
1110	D	D	D	D	D	D	D	A	Vdd	Vss
1111	D	D	D	D	Vref+	Vref−	D	A	RA3	RA2

Figure 2.17 ADCON1 bit definitions

PIC microcontrollers have built-in oscillator circuits and this oscillator can be operated in one of five modes.

- LP – Low-power crystal
- XT – Crystal/resonator
- HS – High-speed crystal/resonator
- RC resistor – capacitor
- No external components (only on some PIC microcontrollers).

In LP, XT, or HS modes, an external oscillator can be connected to the OSC1 input as shown in Figure 2.18. This can be a crystal-based oscillator, or simple logic gates can be used to design an oscillator circuit.

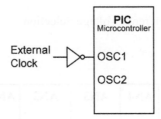

Figure 2.18   Using an external oscillator

Crystal operation

As shown in Figure 2.19, in this mode of operation an external crystal and two capacitors are connected to the OSC1 and OSC2 inputs of the microcontroller. The capacitors should be chosen as in Table 2.6. For example, with a crystal frequency of 4 MHz, two 22 pF capacitors can be used.

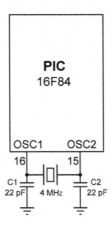

Figure 2.19   Crystal oscillator circuit

Table 2.6 Capacitor selection for crystal operation

Mode	Frequency	C1, C2
LP	32 kHz	68–100 pF
LP	200 kHz	15–33 pF
XT	100 kHz	100–150 pF
XT	2 MHz	15–33 pF
XT	4 MHz	15–33 pF
HS	4 MHz	15–33 pF
HS	10 MHz	15–33 pF

Resonator operation

Resonators are available from 4 to about 8 MHz. They are not as accurate as crystal-based oscillators. Resonators are usually 3-pin devices and the two pins at either sides are connected to OSC1 and OSC2 inputs of the microcontroller. The middle pin is connected to the ground. Figure 2.20 shows how a resonator can be used in a PIC microcontroller circuit.

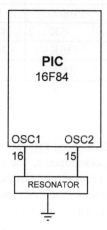

Figure 2.20 Resonator oscillator circuit

RC oscillator

For applications where the timing accuracy is not important we can connect an external resistor and a capacitor to the OSC1 input of the microcontroller as in Figure 2.21. The oscillator frequency depends upon the values of the resistor and capacitor (see Table 2.7), the supply voltage, and to the temperature. For most applications, using a 5 K resistor with a 20 pF capacitor gives about 4 MHz and this may be acceptable in non-time critical applications.

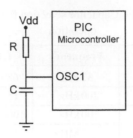

Figure 2.21   RC oscillator circuit

Table 2.7   RC oscillator component selection

C	R	Frequency
20 pF	5 K	4.61 MHz
	10 K	2.66 MHz
	100 K	311 kHz
100 pF	5 K	1.34 MHz
	10 K	756 kHz
	100 K	82.8 kHz
300 pF	5 K	428 kHz
	10 K	243 kHz
	100 K	26.2 kHz

Internal oscillator

Some PIC microcontrollers (e.g. PIC12C672 and PIC16F628) have built-in oscillator circuits and they do not require any external timing components. The built-in oscillator is usually set to operate at 4 MHz and is selected during the programming of the device. For example, the PIC16F62X series of PIC microcontrollers can be operated with an internal resistor– capacitor-based 4 MHz oscillator (called mode INTRC). Additionally, a single resistor can be connected to pin RA7 of the microcontroller to create a variable oscillator frequency (called ER mode). For example, in the PIC16F62X microcontroller OSC1 and OSC2 pins are shared with the RA7 and RA6 pins respectively. The internal oscillator frequency can be set by connecting a resistor to pin RA7 as shown in Figure 2.22. Depending on the value of this resistance the oscillator frequency can be selected from 200 kHz to 10.4 MHz (see Table 2.8). When used in this mode, pin RA7 is not available as a digital I/O pin.

The internal oscillator frequency of some microcontrollers (e.g. PIC16F630) can be calibrated so that more accurate timing pulses can be generated in time critical applications (e.g. in serial communications). In these microcontrollers an oscillator register called OSCCAL is used for the

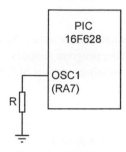

Figure 2.22   Changing the internal oscillator frequency

Table 2.8   Resistor value for the internal oscillator

Resistance	Frequency
0	10.4 MHz
1 K	10.0 MHz
10 K	7.4 MHz
20 K	5.3 MHz
47 K	3 MHz
100 K	1.6 MHz
220 K	800 kHz
470 K	300 kHz
1 M	200 kHz

calibration of the oscillator frequency. A factory-calibrated oscillator constant is loaded into the last location of the memory. By copying this constant value into the oscillator register we can have a more accurate 4 MHz clock frequency for our microcontroller. It is also possible to modify the OSCCAL register values in order to fine-tune the oscillator frequency.

The following PicBasic and PicBasic Pro statements can be used to copy the oscillator calibration constant from the last memory location into the OSCCAL register. These commands must be declared at the beginning of our programs.

```
DEFINE OSCCAL_1 K 1        For 1 K core-size microcontrollers
DEFINE OSCCAL_2 K 1        For 2 K core-size microcontrollers
```

Note that the oscillator constant can be erased during the erasing of the program memory. You should make a note of the value at the last location of the program memory before erasing the memory. If this value is known it can be loaded directly into the OSCCAL register at the beginning of our programs as shown below (here it is assumed that the constant is $24).

```
OSCCAL = $24
```

2.4.5   *Reset circuit*

Reset is used to put the microcontroller into a known state. Normally when a PIC microcontroller is reset execution starts from address 0 of the program memory. This is where the first executable user program resides. The reset action also initialises various SFR registers inside the microcontroller.

PIC microcontrollers can be reset when one of the following conditions occur:

- Reset during power on (POR – Power On Reset)
- Reset by lowering MCLR input to logic 0
- Reset when the watchdog overflows.

As shown in Figure 2.23, a PIC microcontroller is normally reset when power is applied to the chip and when the MCLR input is tied to the supply voltage through a 4.7 K resistor.

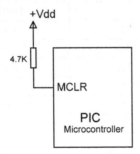

Figure 2.23   Using the power on reset

There are many applications where we want to reset the microcontroller, e.g. by pressing an external button. The simplest circuit to achieve an external reset is shown in Figure 2.24. In this circuit, the MCLR input is normally at logic 1 and the microcontroller is operating normally. When the reset button is pressed this pin goes to logic 0 and the microcontroller is reset. When the reset button is released the microcontroller starts executing from address 0 of the program memory.

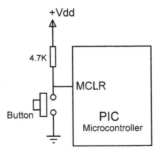

Figure 2.24   Using an external reset button

2.4.6   Interrupts

Interrupts are an important feature of all microcontrollers. An interrupt can either occur asynchronously or synchronously. Asynchronous interrupts are usually external events which interrupt the microcontroller and request service. For example, pin INT (RB0) of a PIC16F84 microcontroller is the external interrupt pin and this pin can be used to interrupt the microcontroller asynchronously, i.e. the interrupt can occur at any time independent of the program being executed inside the microcontroller. Synchronous interrupts are usually timer interrupts, such as the timer overflow generating an interrupt.

Depending on the model used, different PIC microcontrollers may have different number of interrupt sources. For example, PIC16F84 microcontroller has the following four sources of interrupts:

- External interrupt from INT (RB0) pin
- TMR0 interrupt caused by timer overflow
- External interrupt when the state of RB4, RB5, RB6, or RB7 pins change
- Termination of writing data to the EEPROM.

Interrupts are enabled and disabled by the INTCON register. Each interrupt source has two bits to control it. One enables interrupts, the other one detects when an interrupt occurs. There is a common bit called GIE (see INTCON register bit definitions) which can be used to disable all sources of interrupts.

The INTCON control bits of various interrupt sources are

Interrupt Source	Enabled by	Completion Status
External interrupt from INT	INTE = 1	INTF = 1
TMR0 interrupt	T0IE = 1	T0IF = 1
RB4–RB7 state change	RBIE = 1	RBIF = 1
EEPROM write complete	EEIE = 1	–

Whenever an interrupt occurs the microcontroller jumps to the ISR. On low-end microcontrollers (e.g. PIC16F84 or PIC16F628) all interrupt sources use address 4 in program memory as the start of the ISR. Because all interrupts use the same ISR address we have to check the interrupt completion status to detect which interrupt has occurred when multiple interrupts are enabled.

The completion status has to be cleared to zero if we want the same interrupt source to be able to interrupt again.

Assuming that we wish to use the external interrupt (INT) input, and interrupts should be accepted on the low to high transition of the INT pin, the steps before and after an interrupt are summarised below.

- Set the direction of the external interrupt to be on rising edge by setting INTEDG = 1 in register OPTION_REG.

- Enable INT interrupts by setting INTE = 1 in register INTCON.
- Enable global interrupts by setting GIE = 1 in register INTCON.
- Carry out normal processing. When interrupt occurs program will jump to the ISR.
- Carry out the required tasks in the ISR routine.
- At the end of the ISR, re-enable the INT interrupts by clearing INTF = 0.

As we shall see in the projects section of the book, PicBasic Pro language has special instructions for handling interrupts.

2.4.7   The configuration word

PIC microcontrollers have a special register called the *Configuration Word*. This is a 14-bit register and is mapped in program memory 2007 (hexadecimal). This address is beyond the user program-memory space and cannot be directly accessed in a program. This register can be accessed during the programming of the microcontroller.

The configuration word stores the following information about a PIC microcontroller:

- Code protection bits: These bits are used to protect blocks of memory so that they cannot be read.
- Power-on timer enable bit.
- Watchdog (WDT) timer enable bit.
- Oscillator selection bits: The oscillator can be selected as XT, HS, LP, RC, or internal (if supported by the microcontroller).

For example, in a typical application we can have the following configuration word selection during the programming of the microcontroller:

- Code protection OFF
- XT oscillator selection
- WDT disabled
- Power-up timer enables.

2.4.8   I/O interface

A PIC microcontroller port can source and sink 25 mA of current. When sourcing current, the current is flowing out of the port pin, and when sinking current, the current is flowing into the pin. When the pin is sourcing current, one pin of the load is connected to the microcontroller port and the other pin to the ground (see Figure 2.25a). The load is then energised when the port output is at logic 1. When the pin is sinking current, one pin of the load is connected to the supply voltage and the other pin to the output of the port (see Figure 2.25b). The load is then energised when the port output is at logic 0.

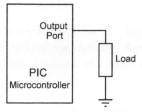

Figure 2.25a   Current sourcing

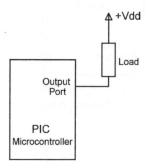

Figure 2.25b   Current sinking

Some useful interface circuits are given in this section.

LED interface

LEDs come in many different sizes, shapes, and colours. The brightness of an LED depends on the current through the device. Some small LEDs operate with only a few milliamperes of current, while standard size LEDs consume about 10 mA of current for normal brightness. Some very bright LEDs consume 15–20 mA of current. The voltage drop across an LED is about 2 V, but the voltage at the output of a microcontroller port is about 5 V when the port is at logic 1 level. As a result of this it is not possible to connect an LED directly to a microcontroller output port. What is required is a resistor to limit the current in the circuit.

If the output voltage of the port is 5 V and the voltage drop across the LED is 2 V, we need to drop 3 V across the resistor. If we assume that the current through the LED is 10 mA, we can calculate the value of the required resistor as

$$R = \frac{5 - 2\,\text{V}}{10\,\text{mA}} = \frac{3\,\text{V}}{10\,\text{mA}} = 0.3\,\text{K}$$

The nearest physical resistor we can use is 330 Ω. Figure 2.26 shows how an LED can be connected to an output port pin in current source mode. In this circuit the LED will be ON when the port output is set to logic 1. Similarly, Figure 2.27 shows how an LED can be connected to an output port pin in current sink mode. In this circuit the LED will be ON when the port output is at logic 0.

Higher current load interface

The circuits given in Figures 2.26 and 2.27 work fine for an LED, or for any other device whose current requirement is less than 25 mA. What do we do if we wish to operate a load with a higher current rating? e.g. a 12 V filament lamp. The answer is that we have to use a switching device, e.g. a transistor or a relay.

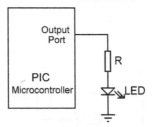

Figure 2.26   Connecting an LED in current source mode

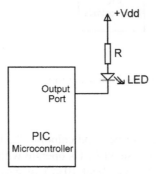

Figure 2.27   Connecting an LED in current sink mode

Figure 2.28 shows how we can drive a small lamp from our port pin using a bipolar transistor. In this circuit, when the port output pin is at logic 1, current flows through the resistor and turns the transistor ON, effectively connecting the bottom end of the lamp to ground. It is important to realise that the positive supply to the lamp is not related to the PIC supply voltage and while the PIC is operating from +5 V, the lamp can be operated from a 12 V supply. The current capability depends upon the type of transistor used and several hundred milliamperes can be achieved with any type of small npn transistors. For higher currents, bipolar power transistors, or preferably MOSFET transistors can be used.

Relay interface

When we want to switch inductive loads such as relays we have to use a diode in the circuit to prevent the transistor from being damaged (see Figure 2.29). An inductive load can generate a back EMF which could easily damage a transistor. By connecting a diode in reverse bias mode this back EMF is dissipated without damaging the transistor.

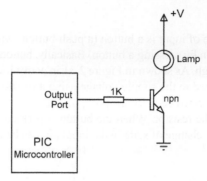

Figure 2.28 Driving a lamp using a transistor

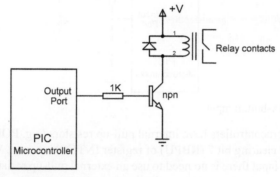

Figure 2.29 Driving an inductive load (e.g. a relay)

Since we can drive a relay, we can connect any load to the relay outputs as long as we do not exceed the contact ratings of the relay. Figure 2.30 shows how a mains lamp can be operated from the microcontroller output port using a relay. The relay could also be operated using a MOSFET power transistor. In this circuit the mains lamp will turn ON when the output port of the micro-controller is a logic 1.

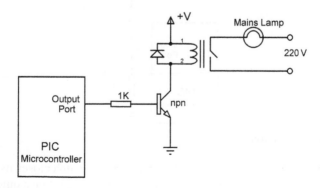

Figure 2.30 Driving a mains bulb using a relay

Button input

One of the most common type of input is a button (a push-button switch) input where the user can change the state of an input pin by pressing a button. Basically, button input can be in two different ways: active low and active high. As shown in Figure 2.31 in active low implementation, the microcontroller input pin is connected to the supply voltage using a resistor (this is also called a pull-up resistor) and the button is connected between the port pin and ground. Normally the microcontroller input is pulled to logic 1 by the resistor. When the button is pressed the input is forced to ground potential which is logic 0. The change of state in the input pin can be determined by a program.

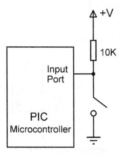

Figure 2.31   Active low-button input

Some ports in PIC microcontrollers have internal pull-up resistors (e.g. PORTB) and these resistors can be enabled by clearing bit 7 (RBPU) of register INTCON to zero. When one of this port pins is used for button input there is no need to use an external pull-up resistor and the button can simply be connected between the port pin and ground.

A button can also be connected in active high mode as shown in Figure 2.32. In this configuration the button is connected between the supply voltage and the port pin. A resistor (this is also called a pull-down resistor) is connected between the port pin and ground. Normally, the port pin is at logic 0. When the button is pressed the port pin goes to the supply voltage which is logic 1.

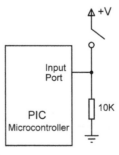

Figure 2.32   Active high-button input

One of the problems with mechanical switches is that when a switch closes its metal parts compress and relax and this causes the switch to open and close several times quickly. The problem is that the microcontroller can read the switch so fast that it can see the switch open and close during the bouncing of the metal parts and this may cause wrong switch state to be read by the microcontroller. One way to eliminate this switch-bouncing problem is to delay reading the input after the switch state changes. For example, when we detect the switch is pressed, we may wait about 10 ms before we read the state of the switch.

In Figures 2.31 and 2.32, we have seen how simple buttons can be connected to a microcontroller port. It is also possible to connect to an input pin a switching transistor, the output of another IC, or simply the output of another PIC port pin. Figure 2.33 shows how a switching transistor can be connected as an input. In this circuit the transistor acts like an inverting switch. When the transistor input voltage is 0 V, the transistor is in OFF state and the port pin is at logic 1 level. When the transistor input voltage is +5 V the transistor turns ON and its collector-emitter voltage drops to 0 V, making the port pin logic 0. One thing nice about this circuit is that the transistor input voltage does not need to be +5 V to turn the transistor ON, it could easily be 9 or 12 V.

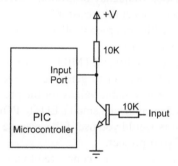

Figure 2.33   Transistor input

The input ports of PIC microcontrollers are protected by internal diodes for over-voltage and under-voltage. Thus, the voltage on a pin can exceed the supply voltage, or it can go below the ground voltage without causing any harm to the microcontroller. The RS232 serial communication lines operate with ±12 V and we can usually connect these lines directly to the input ports using resistors without damaging the microcontroller.

2.5   Exercises

1. What is a flash memory? Explain the differences between a flash program memory and an EPROM program memory. Which one would you use in program development?
2. What is an EEPROM memory? Explain where you might use it. Give an example PIC microcontroller which has EEPROM memory.
3. Explain briefly the bit definitions of the INTCON register. Where would you use bit 6 of this register?
4. Explain how an I/O port direction is controlled in a PIC microcontroller. In an application it is required to make bits 0, 2, 4, and 6 of PORTB as input ports. What value would you have to load into the TRIS register?

5. In an application it is required to make all PORTB pins as inputs and all PORT C pins as outputs. What value would you load into the TRIS registers?

6. Explain what registers are used to control the A/D on a PIC microcontroller. What are the ADRESH and ADRESL registers?

7. In an application it is required to have 3 digital ports and 5 analogue ports. What value would you have to load into register ADCON1?

8. Explain how you can connect an external crystal to a PIC microcontroller. What capacitor values would you choose for a 10 MHz crystal?

9. What are the advantages of using a resonator instead of a crystal?

10. In a simple application where the timing accuracy is not important it is required to operate a PIC microcontroller with a clock frequency of around 2 MHz. What value of resistor and capacitor would you use in the timing circuit?

11. Explain how the internal oscillator can be used on a PIC16F628 microcontroller. It is required to use an internal clock frequency of around 3 MHz. What value of resistor would you use and where would you connect this resistor?

12. Explain what happens when a PIC microcontroller is reset. How can you achieve the reset action by using external components?

13. Explain the differences between TMR0 and TMR1 of a PIC microcontroller.

14. It is required to load the TMR0 register to generate an overflow in 250 ms. Assuming the clock frequency is 4 MHz, choose suitable values for the prescaler and TMR0.

15. In an application it is required to connect 8 small LEDs to PORTB pins of a PIC16F84 microcontroller. What value resistors would you use if the average current of the LEDs are 2 mA? Draw the circuit diagram of your project.

16. Explain the different ways a button can be connected to a microcontroller input port. What are the advantages of using the internal pull-up resistors? Explain how you can enable the internal pull-up resistors of a PIC microcontroller.

17. Explain how a relay can be connected to the output port of a microcontroller. What are the advantages and disadvantages of using relays?

3
PIC microcontroller project development

In this chapter, we will look at the hardware and software tools required to develop PIC microcontroller-based projects. We begin by looking at the minimum hardware tools required and explain the function of each tool.

3.1 Required hardware tools

A PIC microcontroller is an integrated circuit and as such it is useless unless it is programmed and used properly in an electronic circuit to carry out a certain task. The following hardware tools are normally required before a microcontroller-based project can be developed:

- A desktop or a laptop PC
- PIC microcontroller programmer device
- A solderless breadboard or a similar circuit development board
- PIC microcontroller chip(s) and support components
- Power supply

We shall look at each of these tools in detail now.

3.1.1 PC

One of the most important and perhaps the most expensive tools we need is a PC. This can be a desktop PC or a laptop PC. A laptop PC is preferred as it can be carried around and it provides greater flexibility. The PC must be running one of the current Windows operating systems (e.g. Windows 2000 or Windows XP) and it should be equipped with:

- Hard disk with several Giga-byte free space
- CDROM reader
- Floppy drive
- USB port (see notes in later sections)
- Parallel port (see notes in later sections)

Among other things, such as perhaps the *Microsoft Office*, *Internet Explorer*, *Games*, etc., the hard disk will be required to store:

- A text editor software to develop our programs with
- The PicBasic compiler software

- PIC microcontroller programmer software
- The programs that we develop

Most of the commercial software is nowadays distributed on CDROMs and this is why you will need a CDROM reader on your PC. You will find that some small software may still be distributed on floppies and this is why you may also need a floppy drive.

As we shall see in later sections of this chapter, some microcontroller programmer devices are designed to be interfaced to the parallel port (or the printer port) of the PC, while some newer ones are designed for the USB interface. Depending on the type of programmer device you have, you will need either a parallel port or a USB port on your PC. Most laptop PCs are nowadays equipped with only USB ports. If your programmer requires a serial or a parallel port, you can purchase a device to convert between a serial or a parallel interface and the USB.

3.1.2   PIC microcontroller programmer device

A microcontroller programmer device is a stand-alone unit usually with one or more ZIF (zero-insertion-force) type sockets mounted on it. The device is connected to the PC using either a parallel (or sometimes a serial) cable or by the USB interface. The new programmer devices with the USB interface do not require any external power supply as they are powered from the USB port of the PC they are connected to. The older devices with serial or parallel interfaces require an external mains adaptor for their operation. The size of the ZIF socket determines the types of chips that can be programmed by the device. Some sockets are 40-pin which can be used to program microcontrollers with 40, 24, 20, 18, and 8 pins. Some programmer devices have sockets with only 18 pins and they are designed to program smaller microcontrollers with 18 or less pins.

Figure 3.1 shows a typical PIC microcontroller programmer device based on a USB-type interface. This device is distributed by Forest Electronics Ltd. in UK (website www.fored.co.uk) and is known as the *FED Programmer*. The programmer has a single 40-pin ZIF socket mounted on it. Microcontrollers with 40-pins (e.g. PIC16F877) can be programmed by placing them directly on the socket and closing the handle. Devices with less number of pins (e.g. PIC16F84) are normally placed at the far end of the socket near the handle. The Programmer in Figure 3.1 has the advantage that it can program a very large variety of PIC microcontroller chips. The programmer device is sold for around £99 in UK and includes a USB cable.

A PIC microcontroller programmer device designed to operate with the parallel port is shown in Figure 3.2. This particular device is known as the *EPIC Plus* programmer and it can be purchased from the developers of the PicBasic/Pro compilers (microEngineering Labs Inc.) or from many other electronic component distributors. *EPIC Plus* is a low-cost programmer with an 18-pin socket on the device. There is no ZIF socket on the device and a standard DIL (dual-in-line) socket is provided. The programmer is connected to the parallel port (the printer port) of a PC using a 25-way DB25 type cable. If the parallel port of your PC is connected to the printer, the

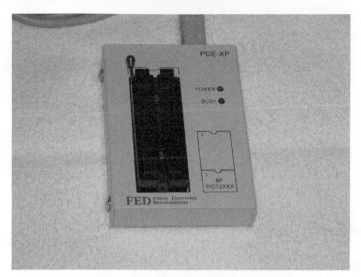

Figure 3.1 USB port-based PIC microcontroller programmer device

Figure 3.2 Parallel port-based PIC microcontroller programmer device

printer must be disconnected while you are using the programmer. *EPIC Plus* is powered from a 12–15 V DC mains adaptor.

Some microcontroller programmer devices have multiple ZIF sockets, also called gang programmers. These programmers are usually used to copy the same program to a number of devices at the same time, such as during the production runs. An example multiple socket programmer is shown in Figure 3.3.

Figure 3.3   Multiple socket programmer (Courtesy of Dataman)

3.1.3   *Solderless breadboard*

When we are building an electronic circuit, we have to connect the components as outlined in the given circuit diagram. This task can usually be carried out on a strip-board or a printed circuit board (PCB) by soldering the components together. The PCB approach is used for circuits which have been tested and which function as desired and also when the circuit is to be made permanent. It is not economical to use a PCB for one or only a few applications.

During the development stage of an electronic circuit, it may not be known in advance whether or not the circuit will function correctly when assembled. A solderless breadboard is then usually used to assemble the circuit components together. A typical breadboard is shown in Figure 3.4. The board consists of rows and columns of holes that are spaced so that integrated circuits and other components

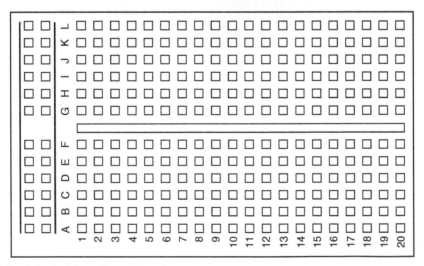

Figure 3.4   A typical breadboard layout

can be fitted inside them. The holes have spring actions so that the component leads can be held tightly inside the holes. There are various types and sizes of breadboards depending on the complexity of the circuit to be built. The boards can be stacked together to make larger boards for very complex circuits. Figure 3.5 shows the internal connection layout of the breadboard given in Figure 3.4.

The top and bottom half parts of the breadboard are separate with no connection between them. Columns 1 to 20 in rows A to F are connected to each other on a column basis. Similarly, rows G to L in columns 1 to 20 are connected to each other on a column basis. Integrated circuits are placed such that the legs on one side are on the top half of the breadboard, and the legs on the other side of the circuit are on the bottom half of the breadboard. The first two columns on the left of the board are usually reserved for the power and earth connections. Connections between the components are usually carried out by using stranded (or solid) wires plugged inside the holes to be connected.

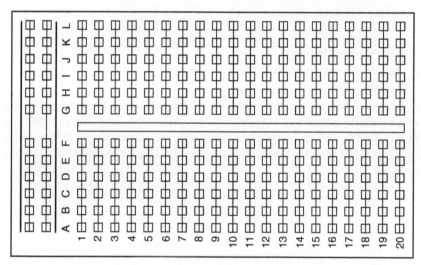

Figure 3.5   Internal wiring of the breadboard in Figure 3.4

Figure 3.6 shows the picture of a breadboard with two integrated circuits and a number of resistors and capacitors placed on it.

The nice thing about the breadboard design is that the circuit can be modified very easily and quickly and different ideas can be tested without having to solder any components. The components can easily be removed and the breadboard can be used for other projects after the circuit has been tested and working satisfactorily.

3.1.4   PIC microcontroller and minimum support components

A PIC microcontroller, even though it may have been programmed, is not of much use unless it is supported by a number of components, such as the timing components and the reset circuitry. As described in Chapter 2, a PIC microcontroller requires an external clock circuit (some PIC

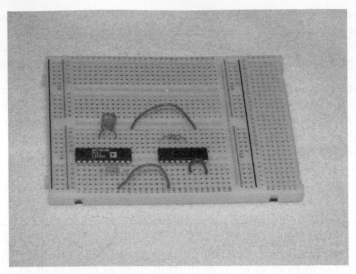

Figure 3.6   Picture of a breadboard with some components

microcontrollers have built-in clock circuits) to function accurately. For accurate timing applications, the clock circuitry consists of a crystal, and two small capacitors. The commonly used crystal frequency is 4-MHz and as described in Chapter 2, the capacitors for this crystal should be around 22 pF. Figure 3.7 shows a 4-MHz crystal with two 22-pF capacitors.

Figure 3.7   A 4 MHz crystal with two 22-pF capacitors

Figure 3.8 shows the circuit diagram of a PIC microcontroller with a 4-MHz crystal clock circuit. The crystal and the capacitors are connected to the OSC1 and OSC2 inputs of the microcontroller.

PIC16F84 microcontroller is taken as an example in all of the figures in this section, but the same principles apply to all the other members of the PIC microcontroller family.

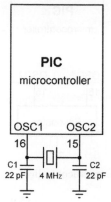

Figure 3.8   PIC microcontroller clock circuit

Resonators are more often used in microcontroller clock circuits because of their low cost, simplicity, and low component count. Figure 3.9 shows some typical 3-terminal resonators, and the connection of a resonator to a PIC microcontroller is shown in Figure 3.10. The centre pin is connected to ground, and the two pins at either sides of the resonator are connected to the OSC1 and OSC2 oscillator inputs of the PIC microcontroller.

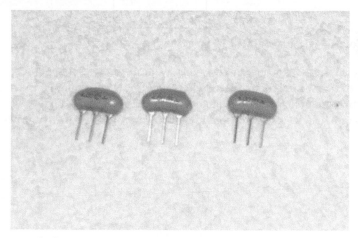

Figure 3.9   Some typical resonators

A PIC microcontroller starts executing the user program from address 0 of the program memory when power is applied to the chip. As shown in Figure 3.11, the reset input (MCLR) of the microcontroller is usually connected to the +V supply voltage through a 4.7K resistor.

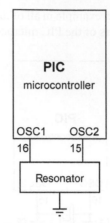

Figure 3.10   Using a resonator in a PIC microcontroller

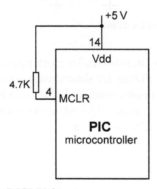

Figure 3.11   Connecting the reset (MCLR) input

There are many applications where the user may want to force reset action e.g. by pressing an external button so that the program re-starts to execute from the beginning. External reset is very useful during microcontroller-based system development and testing. Figure 3.12 shows how an external reset button can be connected to a PIC microcontroller. Normally the MCLR input is at

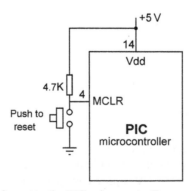

Figure 3.12   Applying external reset to the PIC microcontroller

logic 1, and goes to logic 0 which resets the microcontroller when the reset button is pressed. The microcontroller goes back to the normal operating mode when the button is released.

Now that we have described the clock and the reset circuitry let us look at the design of a minimum PIC microcontroller system. Figure 3.13 shows the circuit diagram of a PIC microcontroller circuit with a 4-MHz resonator and an external reset button. As mentioned earlier, PIC16F84 microcontroller is taken as an example in this figure. The layout of the circuit on a breadboard is shown in Figure 3.14.

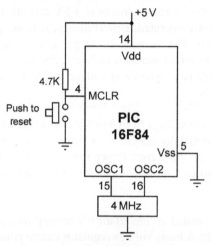

Figure 3.13   Minimum PIC16F84 resonator–based microcontroller circuit

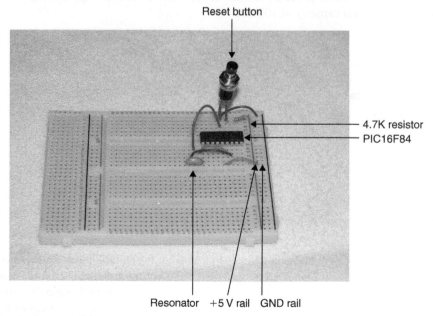

Figure 3.14   Layout of the circuit in Figure 3.13 on a breadboard

3.1.5   Power supply

Every electronic circuit requires a power supply to operate. The required power can either be provided from a battery, or the mains voltage can be used and then reduced to the required level before it is used in the circuit (e.g. a mains adaptor). In this section, we shall look at the design of a power supply circuit to power our PIC microcontroller circuits using a battery as the source of power.

PIC microcontrollers can operate from a power supply voltage in the range 2 to 6 V. The standard power supply voltage in digital electronic circuits is +5 V and this is the voltage with which the PIC microcontrollers are mostly operated. Unfortunately, it is not possible to obtain 5 V using standard alkaline batteries only. The nearest we can get is by using three batteries, which gives 4.5 V and this is not enough to power standard logic circuits. In this section, we shall see how to convert a standard 9-V battery (e.g. type PP3) voltage to 5 V so that it can be used in our PIC microcontroller-based projects.

The simplest solution to drop the voltage from 9 to 5 V is by using a potential divider circuit using two resistors. Although a potential divider circuit is simple, it has the major disadvantage that the voltage at the output depends on the current drawn from the circuit. As a result of this, the output voltage will change as we add or remove components from our circuit. Also, the output voltage falls as the battery is used.

A voltage regulator circuit is needed to convert the 9 V battery voltage into 5 V, independent of the current drawn from the supply. A basic voltage regulator circuit consists of a regulator integrated circuit and filter capacitors. Figure 3.15 shows a low-cost voltage regulator circuit using the 78L05-type voltage regulator IC, and two filter capacitors. 78L05 (see Figure 3.16) is a 3-pin IC with a maximum current capacity of 100 mA.

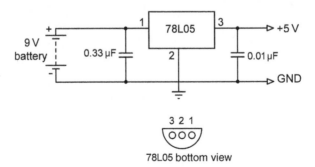

Figure 3.15   Circuit diagram of the +5-V voltage regulator

One of the pins of 78L05 is connected to the +V terminal of the battery in parallel with a 0.33-μF capacitor. One of the pins is connected to the −V terminal of the battery. The third pin provides the +5 V output and a 0.01-μF capacitor should be used in parallel with this pin. In applications where a larger current is required, the 7805 regulator IC can be used. This is pin compatible with

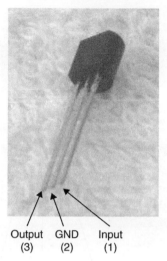

Output   GND   Input
(3)   (2)   (1)

Figure 3.16   78L05 voltage regulator

the low-power 78L05 and it has a maximum current capacity of 1 A. 78L05 should be used with a suitable heatsink in applications drawing more than a few hundreds of milliamperes.

The complete circuit diagram of our PIC16F84-based basic system, together with the power supply, is shown in Figure 3.17. The layout of the circuit on a breadboard is given in Figure 3.18. The circuit in Figure 3.17 is our basic PIC16F84 microcontroller circuit and is now fully functional. What is required now is to write our program and load it into the program memory of the microcontroller. This is the topic of the next chapter.

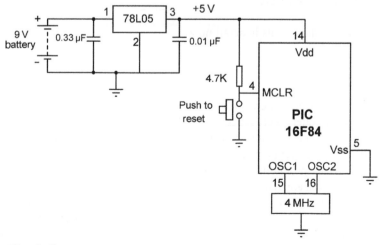

Figure 3.17   Circuit diagram of the complete PIC16F84-based system

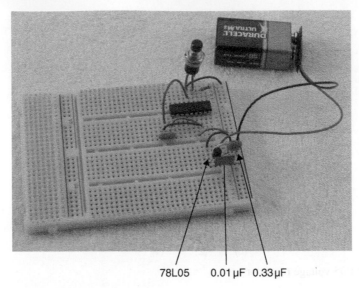

78L05   0.01 μF 0.33 μF

Figure 3.18   Breadboard layout of the system

3.2   Required software tools

All microcontrollers require a program (or software) for their operation. This program is developed and tested by the programmer (or the user). The following software tools are normally required in a PIC microcontroller-based project development cycle:

- A text editor
- *PicBasic* or *PicBasic Pro* compilers
- PIC programmer device software

We shall look at each of these tools in detail now.

3.2.1   Text editor

A text editor helps us write our program (or the source code) so that it can be compiled and loaded into our target microcontroller. There are two text editors readily available on any standard PC – the Windows-based *Notepad*, and the DOS-based *EDIT* (note that *WORD* cannot be used as a text editor since it inserts special control characters into the text). We can use any of these text editors to create a file and write our programs. A program file consists of a file name and a file extension. The file name can be given any name, but the file extension is usually chosen as .*BAS* in PicBasic and PicBasic Pro programs, for example, *MYPROG.BAS* and *LED.BAS* (In general, a program file can be given any other file extension but when the program file is specified when invoking the PicBasic or PicBasic Pro compilers, the file name and file extension must be specified to the compiler. If the file extension .*BAS* is used, then only the file name needs to be specified). It is a good practice to store all of our program files inside a folder so that we can find them easily when we need them.

DOS edit

This is the old PC text editor which runs under the DOS operating system. Although not very powerful, it should be powerful enough to develop small programs. As an example, to create a text file called *LED.BAS* using the DOS editor the steps to follow are:

- Go to the MSDOS prompt. On Windows 2000 and XP machines this is usually found by following the path *START -> Programs -> Accessories -> Command Prompt*. You should then go to the root directory C:\> by entering the commands *CD ..* followed by *CD ..*
- Go to the folder where you want to create your file. If the folder does not exist, create it using the command *MD* followed by the required folder name. For example, to create the folder named MYBASIC, enter the command *MD MYBASIC*. Then move to this folder by entering the DOS command *CD MYBASIC*. You should see the DOS prompt *C:\MYBASIC>* as shown in Figure 3.19.
- Start the DOS editor by entering the command *EDIT LED.BAS*. Write your program using the PicBasic/Pro commands as explained in Chapter 4 and then save the program by pressing the keys *Alt F* and then *X* and then press the *RETURN* key. Your program will be named *LED.BAS* and will be saved inside the folder MYBASIC under the root directory.

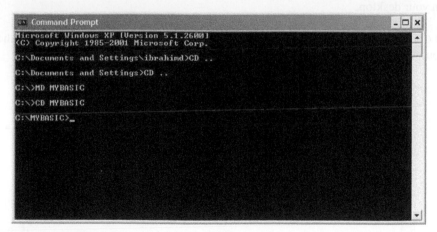

Figure 3.19   Creating the folder MYBASIC in DOS

WINDOWS notepad

Notepad is a powerful text editor which runs under all Windows operating systems. Notepad can be accessed by following the path *START -> Programs -> Accessories -> Notepad*. You should write your programs and then save them with the *.BAS* file extensions. It is important that when you save the file you should select the *Save As Type* as *All Files* in the Notepad File save menu.

Although both Edit and Notepad are useful for creating new programs or for modifying an existing program, Integrated Development Environments (IDE) such as the *CodeDesigner* and *MicroCode Studio* provide a much easier and quicker method of program development since they combine a powerful syntax highlighted editor with the compiler and the device programmer software. As a result, the programmer can develop the program, compile it, and then load it into the target micro-controller by using only one program interface. Both of these products are third-party products and can be purchased from the developers, or in some cases, cut-down versions can be downloaded free of charge from the Internet. We shall be looking at both products in the following sections.

CodeDesigner

The CodeDesigner software package has been developed by CSMicro Systems (web site www. csmicrosystems.com). A cut-down version of the CodeDesigner, known as CodeDesigner Lite can be downloaded free of charge from the microEngineering Labs Inc. web site (www.melabs.com). The installation and configuration instructions are also available from their web site. After downloading the software, double click the cdlite icon and then follow the standard Windows installation procedures. When the installation is finished, click the Finish button. The software is installed in the root directory inside folder C:\CDLITE>. Now, create a shortcut to CodeDesigner Lite in your desktop so that you can invoke the program easily. To do this, open the Windows Explorer and navigate to My Computer -> Local Disk (C:) and click on Local Disk(C:). Then click on folder CDLite. Find application cdlite on the right hand window and right click on the application. Then select Send To -> Desktop (create shortcut). You will now have a shortcut named Shortcut to cdlite in your desktop.

To start *CodeDesigner*, double-click on the shortcut you have just created. Figure 3.20 shows the form you will see on your screen.

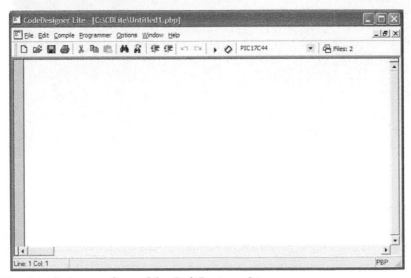

Figure 3.20   A typical screen form of the *CodeDesigner Lite*

CodeDesigner Lite should be configured before it is used. Configuration involves specifying the compiler and the programmer devices to be used in project development. To configure the compiler options, select *Compile -> Compiler* Options from the top menu. If you are using the PicBasic compiler, specify the path to the compiler as shown in Figure 3.21 and press OK.

If you are using the PicBasic Pro compiler, specify the path to the compiler as shown in Figure 3.22.

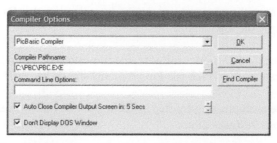

Figure 3.21 Configuring *CodeDesigner Lite* for the PicBasic compiler

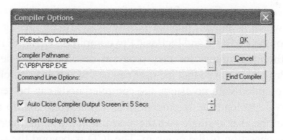

Figure 3.22 Configuring CodeDesigner Lite for the PicBasic Pro compiler

The *CodeDesigner Lite* software should then be configured for the PIC microcontroller programmer device you are using. To do this, select *Programmer -> Programmer Options* from the top menu. You can now choose your programmer device from the given list. If your programmer is not specified in the list, choose *Other* and specify the path to your programmer application software. In this book, we shall be using the FED Programmer shown in Figure 3.1. Figure 3.23

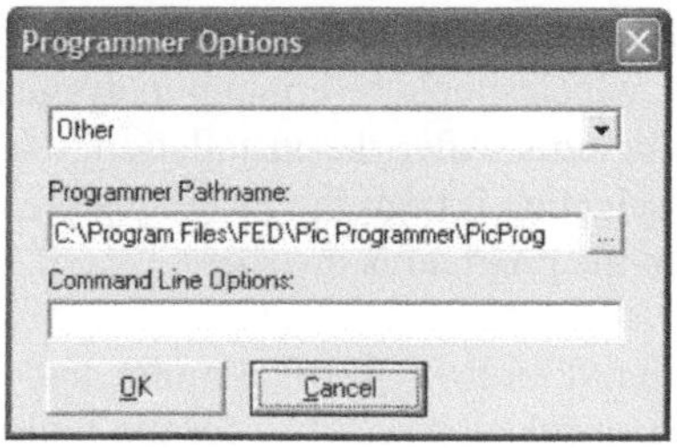

Figure 3.23 Specifying the path to the FED Programmer

shows how to specify the path to this programmer inside the *CodeDesigner Lite* (if you are using a different programmer device then you should either select your device from the list if it is available, or choose *Other* and enter the path to your programmer device software).

The *CodeDesigner Lite* is now ready for program development, compilation, and downloading the code to the target PIC microcontroller. After writing our program, we can choose *Compile ->Compile* from the top menu to compile our program. If the compilation is successful, we can download our program to the programmer device by selecting the *Programmer -> Launch Programmer* options from the top menu.

Note that when using the *CodeDesigner* software, the file extension of PicBasic programs should be *.PBC*, and the file extension of PicBasic Pro programs should be *.PBP*.

We shall see a complete example, step-by-step in Section 3.4 on how to create a project from first principles using the *CodeDesigner Lite*.

MicroCode studio

Although *CodeDesigner Lite* is sufficient for most of our project development tasks, we shall look at *MicroCode Studio,* which is another popular *IDE* with In Circuit Debugging (ICD) capability, designed specifically for the PicBasic and PicBasic Pro compilers. This IDE also provides a syntax-highlighted editor to the programmer for easy program development. The IDE is interfaced to PicBasic or PicBasic Pro compilers so that the user can easily and very quickly compile programs. After the program is compiled with no errors, the compiled code can be sent to a PIC microcontroller programmer device to load the microcontroller. *MicroCode Studio* also provides an ICD capability which enables the user to single-step the program in the target microcontroller in order to examine and verify the operation of the program. The ICD is beyond the scope of this book and interested readers are referred to the manufacturers' web site at *www.mecanique.co.uk*.

MicroCode Studio can be downloaded from the manufacturers' web site and it is available free of charge to non-commercial users. The software is a cut-down version of the full product *MicroCode Studio Plus* but it can be used in all of the projects developed in this book.

MicroCode Studio is also distributed free of charge and is installed as part of the PicBasic Pro compiler demo package from microEngineering Labs Inc. As we shall see in the next section, this package enables the user to create limited programs with a maximum line count of 31 (excluding comment lines and blank lines), which should be enough to evaluate the compiler and to develop many small to medium-size programs. After the installation, *MicroCode Studio* is invoked by double clicking on its icon (or selecting it from the Programs menu) and the screen form shown in Figure 3.24 is displayed when the program is invoked.

The software needs to be configured for the type of compiler, and the type of programmer we are using. When the software is first invoked, it searches for the PicBasic compiler on the hard disk and the compiler path is set automatically if the compiler is found. If the compiler is not found we

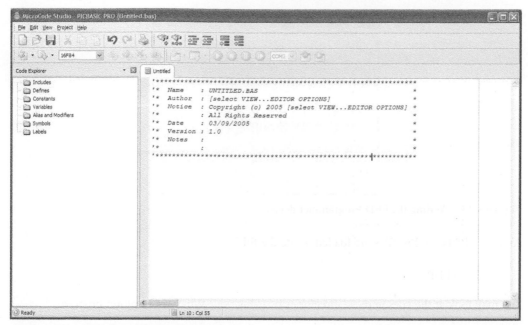

Figure 3.24   *MicroCode Studio* screen form

can specify the path to the compiler by selecting *View -> Compile and Program Options*. Then select the *Compiler* tab and specify the compiler path by clicking the *Find Manually* option. You can also click the *Find Automatically* button to see if the compiler path can be found automatically. The type of programmer device we are using should be configured by choosing *View -> Compile and Program Options* and then clicking the *Programmer* Tab. Depending on the type of programmer device we have, we can either choose the default one or choose *Add New Programmer* to add our own programmer device. Figure 3.25 shows how the FED Programmer device can be selected to be the default programmer.

We can now write our program and when finished, compile it by selecting *Project -> Compile* or we can send the code to a PIC programmer by selecting *Project -> Program*.

3.2.2   PicBasic and PicBasic Pro compilers

These compilers are distributed on a floppy diskette or on a CDROM and they should be installed before they can be used. The installation is very easy – insert the diskette into drive A and click *START -> Run* and type *A:\INSTALL* in the RUN dialoug box. The compiler files will automatically be loaded onto the hard disk. PIC Basic files are loaded inside the folder

C:\> PBC

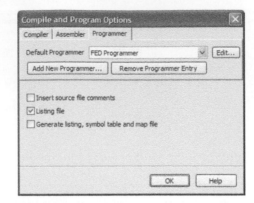

Figure 3.25   Adding the FED Programmer device

And the PIC Basic Pro files are loaded inside the folder

C:\> PBP

You may look at the files in these directories by using your Windows Explorer program.

The compilers can either be activated directly from DOS, or by using *CodeDesigner* or the *MicroCode Studio* as described in Section 3.2.1.

To activate the compilers directly from DOS, go to the *Command Prompt* mode and then enter

C:\PBC> PBC –pxxx myfile   for PicBasic compiler

and

C:\PBP> PBP –pxxxx myfile   for PicBasic Pro compiler

where

–pxxx is the PIC microcontroller type (e.g. -p16F877 for PIC16F877). If the microcontroller type is not specified, the default PIC16F84 is assumed;
–myfile is the name of the program to be compiled (a .BAS file extension is assumed).

For example, the following command assumes that we are using a PIC16F627 microcontroller, and compiles PicBasic Pro file called *LED.BAS*. The file is assumed to be in the same directory as the compiler:

C:\PBP> PBP –p16F627 LED

Similarly, the following command can be used to compile a program called *MOTOR.BAS* using the PicBasic compiler. It is assumed here that our target system is a PIC16F84 microcontroller.

C:\PBC> PBC MOTOR

A demo version of the PicBasic Pro compiler is available from the web site of microEngineering Labs Inc. and this is included on the CDROM distributed with this book. You can use this demo version to create programs with up to 31 lines long. The demo version also includes the *MicroCode Studio* which can be installed during the installation of the compiler.

The compiler generates a number of files with the same filename but with different extensions (for example, .ASM, .HEX etc). The file with the extension .HEX is also known as the object file and this is the file which is to be sent to the programmer device.

3.2.3   Programmer device software

You should install the programmer software which has been distributed with your programmer device. In this book, the USB-based FED Programmer device is used and the software for this device is installed by following the standard Windows software installation procedures. The programmer software is invoked automatically when working with *CodeDesigner Lite* or with the *MicroCode Studio*. Figure 3.26 shows the typical screen form of the FED Programmer software. First of all, you should select the type of PIC microcontroller you will be using. To do this, click *PIC* from the top menu and then click *Select Device* (see Figure 3.27) and select your microcontroller from the given list. The device name you have chosen should appear at the bottom left-hand corner of the screen form.

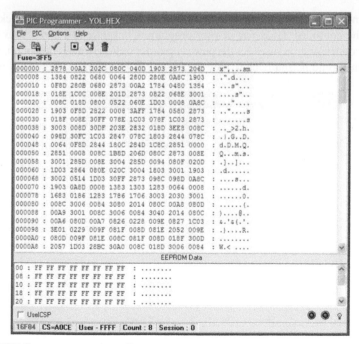

Figure 3.26   FED Programmer screen form

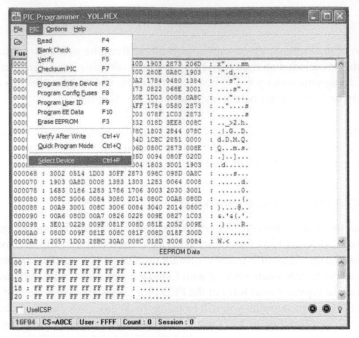

Figure 3.27   Selecting a PIC microcontroller

Insert the PIC microcontroller chip into the socket and close the handle. Then, click *File* and then *Open* to open the compiled .HEX file of your program. Click *File* and then *Fuses* to set the PIC microcontroller configuration fuses for the power-up timer option, watchdog option, and the timing device used. You should normally click only the crystal (XT) option as shown in Figure 3.28. You

Figure 3.28   Setting the configuration fuses

can now program the configuration fuses by selecting *PIC* followed by *Program Config Fuses*. The microcontroller can then be programmed by selecting *PIC* followed by *Program Entire Device*.

3.3 Bundled development systems

Some manufacturers provide bundled packages of their hardware and software products mainly for development and experimenting purposes. Bundled packages have the following advantages:

- The cost is lower than the cost of purchasing the individual products in the package.
- They usually contain all the necessary hardware for developing microcontroller-based products.
- They usually contain the compiler software and programmer software to enable the user develop projects easily.

Some bundled packages for PIC microcontrollers, including the PicBasic or PicBasic Pro compilers are described in this section. All of the bundled products given in this section are manufactured by microEngineering Labs Inc. Further information can be obtained from their web site www.melabs.com

Developer's bundle

This is a complete PIC microcontroller project development package and as shown in Figure 3.29, the package contains

- PicBasic Pro compiler
- Melabs serial programmer device
- LAB-X1 Experimenter board
- PIC microcontroller chips
- All the necessary mains adaptors and interface cables

Figure 3.29 Developer's Bundle (Courtesy of microEngineering Labs Inc.)

PicBasic compiler bundle

This package is based on the PicBasic compiler. The package contains (see Figure 3.30)

- PicBasic compiler
- EPIC Plus programmer
- PICPROTO18 Experimenter board
- PIC microcontroller chips
- All the necessary mains adaptors and interface cables

Figure 3.30   PicBasic Compiler Bundle (Courtesy of microEngineering Labs Inc.)

LAB-X1 bundle with serial programmer

This bundle is for those people who have the PicBasic or the PicBasic Pro compilers and are looking for a programmer device and an experimenter board. The package contains (see Figure 3.31)

- LAB-X1 Experimenter board
- Melabs serial programmer
- PIC microcontroller chips
- All the necessary mains adaptors and interface cables

Fig. 3.31   LAB-X1 Bundle with serial programmer (Courtesy of microEngineering Labs Inc.)

PicBasic or the PicBasic Pro compilers can be added to the bundle at a reduced cost.

3.4   Experimenter boards

In Section 3.1.3, we have seen how to use a solderless breadboard to develop microcontroller-based projects easily and also at low cost. Some manufacturers provide experimenter boards for the development and testing of microcontroller-based systems. Some low-cost experimenter boards contain LEDs and push-button switches. Some more expensive ones may contain LCD displays, keyboards, serial input/output ports, relays, on-board chip programmers, and so on. Examples of some popular experimenter boards are given below.

LAB-X1 experimenter board

This board is manufactured by the microEngineering Labs Inc. Some of the features of this board are (see Figure 3.32)

- A keypad with 16 switches
- Potentiometers, IR, real-time clock
- LED bargraph
- LCD module
- RC servo connectors
- Speaker
- RS232 and RS485 interface
- Serial EEPROM
- Prototyping area
- 5-V regulator

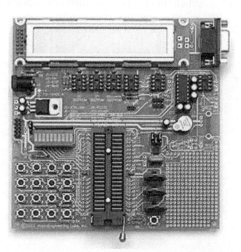

Figure 3.32   LAB-X1 Experimenter board (Courtesy of microEngineering Labs Inc.)

The company also manufactures other experimenter boards such as LAB-X2, LAB-X3, LAB-X4, and so on.

PIC microcontroller training and development kit

This board is manufactured by Kanda Systems Ltd. Some of the features of the board are (see Fig. 3.33)

- A/D converters
- RS232 interface
- 4-digit, 7-segment display
- LED bar-graph
- 8 push-button switches
- Piezo-buzzer
- Infrared transmitter–receiver
- Sockets for serial EEPROM

Figure 3.33   Kanda System's Development kit (Courtesy of www.kanda.com)

EasyPIC 2 development system

This is a very sophisticated development board manufactured by MikroElektronika. The board supports 8, 14, 18, 28, and 40-pin PIC microcontrollers. Some of the important features of the board are (see Figure 3.34)

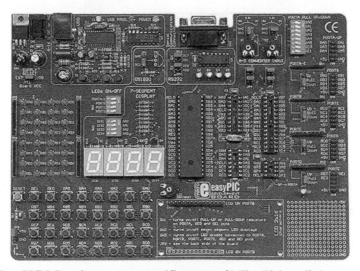

Figure 3.34   EasyPIC 2 Development system (Courtesy of MikroElektronika)

- RS232 interface
- 4-digit, 7-segment display
- 32 push-buttons
- Digital thermometer
- 32 LEDs
- A/D converters
- 2 potentiometers
- On board USB programmer

3.5 Example project development

In this chapter, we have seen the hardware and software tools required to develop a PIC microcontroller–based project. We shall now summarise the steps required for the development of a project by giving a simple example.

In this example, we shall connect a small LED to port RB7 (bit 7 of PORTB) of a PIC16F84 microcontroller and then write a program to continuously flash the LED with 1-s intervals; i.e. the LED will be ON for 1 s, then OFF for 1 s, then ON again for 1 s, and so on. You may have difficulty in understanding the operation of the program given in this section as you may have not read Chapter 4 yet. You should not worry about the details of the actual program since this exercise is not designed to teach you programming, but to show you the steps required for a typical project development cycle.

Step 1 – design the circuit

The circuit diagram of the project is shown in Figure 3.35. A small LED is connected to port RB7 (pin number 13) of a PIC16F84 microcontroller through a current-limiting resistor. The voltage across an LED is about 2 V, and the average current through an LED depends on the type of LED we are using, but we can assume a current of about 10 mA. If we assume that the voltage at the output of an output pin is 5 V, the value of the required current-limiting resistor is then found as

$$R = \frac{5 - 2\,\text{V}}{10\,\text{mA}} = \frac{3\,\text{V}}{10\,\text{mA}} = 0.3\text{K}$$

0.3K is not a standard resistor and we can choose the resistor as $330\,\Omega$ which will give slightly less than 10 mA through the LED.

The microcontroller is operated from a 4-MHz resonator and an external reset button is connected as described in Section 3.1.4. A 9-V battery together with a voltage regulator circuit is used to power the microcontroller as shown in Section 3.1.5.

Step 2 – required components

Make a list of the required components:

- Solderless breadboard
- PIC16F84 microcontroller

- 4-MHz resonator
- Push-button switch
- 4.7K resistor
- LED
- 330-Ω resistor
- 78L05 regulator
- 0.33-μF capacitor
- 0.01-μF capacitor
- 9-V battery clip
- 9-V battery

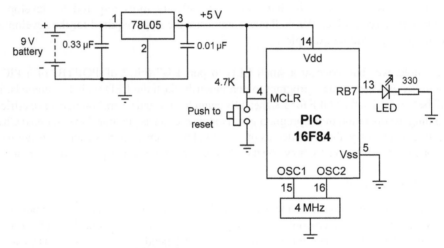

Figure 3.35   Circuit diagram of the project

Step 3 – construct the circuit

Figure 3.36 shows the circuit constructed on a solderless breadboard. You should connect the battery and make the following checks *before* inserting the microcontroller in its place. A voltmeter (e.g. a digital test meter) will be required for these checks.

- Inspect the breadboard visually to make sure that all the connections are correct.
- Measure the voltage at the +5 V rail and make sure that the voltage is very close to +5 V. You should check your battery connections and the 78L05 regulator connections if the voltage is not close to +5 V. You should not continue with the project unless you get the correct voltage at this step.
- Measure the voltage at pin position 14 of the PIC microcontroller chip. Again, this voltage must be very close to +5 V and you should not continue until you get the correct voltage.
- Disconnect the battery

You can now insert the microcontroller chip in its place, but wait until after the chip is programmed.

Figure 3.36 Project constructed on a breadboard

Step 4 – write the program

Before writing our program, let us assume that we shall be keeping all of our programs in a folder named MYPROGS under the root directory. To do this, the following steps will be required (this task will have to be done only once):

- Start the Windows Explorer and click on *My Computer*. Then click on *Local Disk (C:)*. Click on *File* in top menu and then select *New -> Folder*. Enter the name of the new folder as MYPROGS and press the RETURN key. The new folder has now been created and you may Exit the Windows Explorer.

At this part of the development, we shall assume that we are using the PIC Basic Pro compiler.

- Double-click the *CodeDesigner* icon in the Desktop to start the program and make sure that the Compiler Option chosen is the PicBasic Pro.
- Select the microcontroller type as PIC16F84 by clicking on the top middle part of the form, left of the Files:1.
- Write your program by entering the statements shown in Figure 3.37.
- Click on *File* in the top menu and save your program with the name MYLED in the folder MYPROGS (note that the file extension is chosen as .PBP automatically).
- Compile the program by selecting *Compile* from the top menu and then click on *Compile*. Make sure that there are no errors in the compilation.
- Connect the programmer device to your PC and insert a PIC16F84 chip into the programmer device. Click on *Programmer* in the top menu and select *Launch Programmer*. You should now see the programmer software on your screen. Click on *PIC* in the top menu and select the device type as PIC16F84.

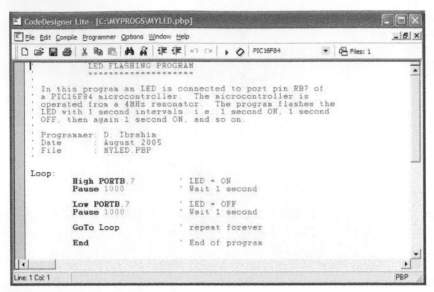

Figure 3.37   The program of our project

- Click on *File* in the top menu and select *Open*. Navigate to folder MYPROGS and click on file MYLED. Click on *OPEN* to load the object file of your program (MYLED.HEX) to the programmer memory (see Figure 3.38).

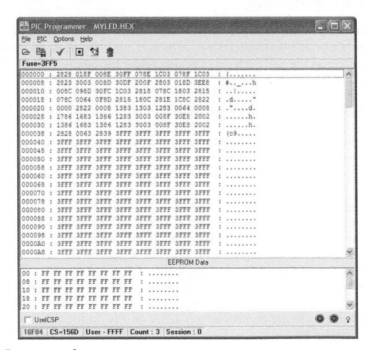

Figure 3.38   Programmer form

- Click on *File* in the top menu and select *Fuses*. In this form, tick only the XT box to indicate that we are using a crystal for timing.
- Click on *PIC* in the top menu and select *Program Config_Fuses* to program the configuration fuses. You should get a confirmation when the configuration fuses have been programmed.
- Click on *PIC* in the top menu and select *Program Entire Device*. Wait until the chip is programmed. You should get a confirmation when the device has been programmed.
- Remove the chip from the programmer and place it on the breadboard. Connect the battery and test your project. The LED should be flashing with 1-s intervals. If the project is not working, first check the hardware to make sure that the connections are correct. Then check the software.

3.6   Other useful development tools

In addition to the microcontroller hardware and software development tools described in this chapter, there are some other development tools which could be very useful during the project development cycle. Two of such tools are described briefly in this section.

3.6.1   Simulators

A simulator is a software development tool designed to run on the PC. A simulator enables the programmer (or the user) to test the functional operation of a program on the PC, without having to construct any microcontroller-based hardware.

Typically, the programmer develops the program and then compiles it. The simulator program is then invoked and the object code of the program is loaded into the simulator program (some simulators have built-in compilers or assemblers which make it easier to write a program, compile it and then simulate using the same development tool). The programmer can then single-step through the program and observe the values of variables as they change. Some simulators provide input–output ports where the programmer can connect various software-simulated devices such as LEDs, 7-segment displays, LCDs, motors, and so on. The programmer is also allowed to change the values of variables during a simulation session so that the operation of the program can be analysed in detail and any logic errors can be removed before the program is loaded into a microcontroller.

Although the simulators can be very useful development tools, they have the disadvantage that the program is not run in real-time. Another disadvantage is that it is not possible to examine the operation of the program when real hardware devices are connected to the input–output ports. For example, it is not possible to connect a real motor to the simulator and see it running. As a result of this, any hardware-related timing errors cannot be detected by the simulation process.

3.6.2   In Circuit Emulators (ICE)

This is another useful microcontroller development tool. In an ICE application, the microcontroller of the target system is replaced by the ICE which behaves exactly same as the original microcontroller. Typically, the microcontroller is removed from its socket and replaced by the ICE header.

This header is usually connected to an emulator box which contains the main emulator functionality. A PC is then connected to the emulator box. The ICE emulates the replaced microcontroller in real-time as if the replaced microcontroller is in the socket. The programmer can load the program he has developed into the emulator and can run, single-step, and trace the operation of the program. Some emulators have advanced functions such as performance analysis, trace buffer, triggering functions, and breakpoint features. Breakpoints give the programmer the ability to stop the program at precise locations and then to examine the values of variables at these points.

A simpler, and also much cheaper type of emulator is an In Circuit Debugger (ICD). ICD provides real-time emulation of the target processor. The program can be executed in single-step mode with breakpoints. Memory locations and values of various registers can be examined in real-time.

3.7   Exercises

1. Describe the minimum hardware tools required to develop PIC microcontroller-based projects.
2. Explain why a PC is needed to develop PIC microcontroller-based projects.
3. What is the function of a programmer device? What types of programmer devices are there?
4. Explain what a breadboard is and why it can be useful during microcontroller-based project development. What are the advantages and disadvantages of using a breadboard?
5. Explain why a power supply is required to power a PIC microcontroller. Draw the circuit diagram of a typical low-cost +5-V power supply.
6. Describe the minimum software tools required for the development of PIC microcontroller-based projects.
7. Explain why you need a text editor. Give examples of at least two text editors available on your PC.
8. Explain what the advantages of using an Integrated Development Environment (IDE) are. Give an example of an IDE for the development of PIC microcontroller-based projects.
9. Explain what *CodeDesigner* is and the advantages of using it.
10. Explain the benefits of using the *MicroCode Studio* software package during the development of PIC microcontroller projects.
11. Explain in detail the steps required to develop a simple PIC microcontroller-based project. Can you suggest some methods to speed-up the development time?
12. Explain where and why you might need to use a simulator. What are the limitations of simulators?

3.8   Links to useful web sites

Links to some useful web sites on PIC microcontrollers and development tools are listed in this section.

microEngineering Labs Inc.	www.melabs.com
MikroElektronika	www.mikroelektronika.co.yu
Kanda Systems Ltd.	www.kanda.com

Maplin Electronics	www.maplin.co.uk
RS Components	www.rswww.com
Farnell In One	www.farnell.com
Mecanique	www.mechanique.co.uk
CSMicro Systems	www.csmicrosystems.com
Brunning Software	http://brunningsoftware.co.uk
Microchip Technology Inc.	www.microchip.com
Images SI Inc.	www.imagesco.com
Microcontroller Pros Corporation	http://microcontrollershop.com
ASIX	www.pic-tools.com
HVW Technologies Inc.	www.hvwtech.com
Microdesigns Inc.	www.microdesignsinc.com
Apogeekits	www.apogeekits.com
Quasar Electronics	www.quasarelectronics.com
Spectro Technologies Inc.	www.spectrotech.net
Dontronics	www.dontronics.com
Hobby Engineering	www.hobbyengineering.com
Mouser	www.mouser.com
ProtoCessor	www.protocessor.com
Crownhill Associates	www.crownhill.co.uk

4

PicBasic and PicBasic Pro programming

BASIC is one of the oldest and one of the easiest programming languages to learn. You should be able to learn and program in BASIC in less than an hour. In this chapter, we shall be looking at the principles of programming PIC microcontrollers using the PicBasic and PicBasic Pro languages. Both these languages are very similar to the standard BASIC language but they have some modified and some additional instructions specifically for microcontroller programming.

Both PicBasic and PicBasic Pro languages have been developed by the microEngineering Labs Inc. PicBasic is a lower-cost, simpler language than PicBasic Pro and it is aimed at students and hobbyists. PicBasic Pro is more expensive, aimed at professionals, and includes additional commands for more advanced instructions.

Table 4.1 gives a list of the comparison of PicBasic and PicBasic Pro languages. Before we proceed to the chapter on PIC applications and projects, we shall be looking at how we can program the PIC microcontrollers using these languages.

4.1 PicBasic language

In this section, we shall be looking at the variable types and the commands of the PicBasic language. A detailed description of all the commands can be found in the PicBasic Compiler manual, available from the web site www.melabs.com, or a printed copy can be obtained from the microEngineering Labs Inc.

4.1.1 PicBasic variables

Variables are used to store temporary data in a program. These variables are stored in the general-purpose area of the RAM memory of a microcontroller.

Variables in PicBasic can be bytes (8 bits), or words (16 bits). Byte variables are named B0, B1, B3, etc., and word variables are named W0, W1, W2, etc. Word variables are made up of two bytes. For example, W0 uses the same memory space as bytes B0 and B1. Similarly, W1 word variable is made up of bytes B2 and B3, and so on. We can access the bit positions of variables B0 and B1 using predefined names Bit0, Bit1,...,Bit15. For example, the least significant bit of B0

Table 4.1 Comparison of PicBasic and PicBasic Pro

PicBasic	PicBasic Pro
Low-cost ($99.95)	Higher cost ($249.95)
Limited to first 2 K of program space	No program space limit
Interrupt service routine in assembly language	Interrupt service routine can be in assembly language or in PicBasic Pro
Peek and Poke used to access registers	Registers can be accessed directly by specifying their names
Some commands can be used only for PORTB, PORTC, or GPIO	Commands can be used for all ports
Clock speed 4 MHz	Any clock speed up to 40 MHz
Most 14-bit Pic microcontrollers supported	All PIC microcontrollers, including 12-bit ones are supported
More code space in memory	5–10% less code space in memory
More difficult to learn and less powerful	Easier to learn and more powerful
No LCD commands	Special LCD control commands (LCDOUT, LCDIN)
No hardware serial communication commands	Special hardware serial communications commands (HSERIN, HSEROUT)
No PWM commands	Special PWM commands for the microcontrollers that have built-in PWM circuit (HPWM)
No Select-Case command	Select-Case command for multi-way selection
No program memory read–write commands	Commands to read and write program memory locations (READCODE, WRITECODE)
No One-wire device interface	One-wire device interface commands (OWIN, OWOUT)
No USB commands	USB commands for microcontrollers that have built-in USB circuits (USBIN, USBOUT)
No X-10 remote control commands	X-10 remote control commands (XIN, XOUT)
No A/D commands	A/D commands for microcontrollers that have built-in A/D converters (ADCIN)

is labelled Bit0, the second bit Bit1, and the most significant bit as Bit7. Similarly, the least significant bit of B1 can be named as Bit8, and the most significant bit of B1 as Bit15.

Variables are stored in the RAM memory of a PIC microcontroller where B0 is the first RAM location, B1 is the second RAM location, and so on. The size of the RAM memory depends on the type of PIC microcontroller used and Table 4.2 gives a list of the variable names for various microcontrollers. For example, if we are using a PIC16F84-type microcontroller, we can define 52 variables from B0 to B51, and the highest variable name must not exceed B51. Note that you can only access RAM locations up to the available RAM. For example, if you try to access a RAM location that does not exist, the compiler does not generate an error and your program may not work as expected.

Table 4.2 PicBasic variable names

Microcontroller	Variables (bytes)	Variables (words)
PIC16C61	B0–B21	W0–W10
PIC16C71	B0–B21	W0–W10
PIC16C710	B0–B21	W0–W10
PIC16F83	B0–B21	W0–W10
PIC16C84	B0–B21	W0–W10
PIC16F83	B0–B21	W0–W10
PIC12F629	B0–B47	W0–W23
PIC12F675	B0–B47	W0–W23
PIC16F630	B0–B47	W0–W23
PIC16F676	B0–B47	W0–W23
PIC16C711	B0–B51	W0–W25
PIC16F84	B0–B51	W0–W25
PIC16C554	B0–B63	W0–W31
PIC16C556	B0–B63	W0–W31
PIC16C620	B0–B63	W0–W31
PIC16C621	B0–B63	W0–W31
PIC 12C67X	B0–B79	W0–W39
PIC14C000	B0–B79	W0–W39
PIC16C558	B0–B79	W0–W39
PIC16C558	B0–B79	W0–W39
PIC16C622	B0–B79	W0–W39
PIC16C62	B0–B79	W0–W39
PIC16C63	B0–B79	W0–W39
PIC16C64	B0–B79	W0–W39
PIC16C65	B0–B79	W0–W39
PIC16C72	B0–B79	W0–W39
PIC16C73A	B0–B79	W0–W39
PIC16C74A	B0–B79	W0–W39

The relationships between the byte, word, and bit variables are given in Table 4.3. For example, word W2 is made up of bytes B4 and B5. You will see additional predefined variables in Table 4.3, named *Port*, *Dirs*, and *Pins*. *Pins* refers to the PORTB hardware, *Dirs* refers to the port data direction register for PORTB, i.e. TRISB and a 0 sets its associated Pin to an input, and a *Dirs* of 1 sets its associated Pin to an output. *Port* is a word variable that combines *Pins* and *Dirs*. The individual pins of a port can be accessed by the variable names Pin0, Pin1,…,Pin7.

Table 4.3 Relationship between byte, word, and bit variables

Word variable	Byte variable	Bit variable
W0	B0	Bit7, Bit6,…Bit0
	B1	Bit15, Bit14,…Bit8
W1	B2	
	B3	
W2	B4	
	B5	
W3	B6	
	B7	
	…	
	…	
W39	B78	
	B79	
Port	Pins	Pin7, Pin6,…Pin0
	Dirs	Dir7, Dir6,…Dir0

Symbols

In order to make programs more readable, we can assign meaningful names to variables, instead of using B0, B1, etc. The PicBasic statement *symbol* is used for this purpose. For example, we can assign variable name count to location B0 with the instruction:

Symbol count = B0

Symbols must be declared at the top of a program. Symbols can also be used to assign constants to names. For example, the following statement assigns the decimal value 20 to the name *total*. Note that this statement does not occupy any location in the microcontroller RAM memory. The number is simply represented with a name.

Symbol total = 20

Command names in PicBasic are case insensitive and can be written in upper case, lower case, or with a mixture of the two. Thus, all the variables below are the same:

TOTAL
Total
toTal

Comments

Comments are useful in programs to describe the operation performed in a line or in a block of lines. A comment starts with either the keyword REM or the single quote character ('). All the characters following a comment character are ignored. Examples of comments are:

```
REM        This is a simple test program
LOW 0      ' Clear Pin 0 to 0
HIGH 1     REM Set Pin 1 to 1
```

Numeric Values

In PicBasic, numeric values can be specified in three ways: decimal, binary, and hexadecimal. Decimal values are the default and require no prefix. Binary values are specified using the prefix "%" followed by the number. Hexadecimal values are specified using the prefix "$" followed by the number. Some examples are:

```
REM A has the same value in all the following three statements
A = 10
A = %00001010
A = $0A
```

ASCII Values

Character constants can be converted into their ASCII values by enclosing them in double quotes. Only one character must be specified. For example,

```
"A"        ' ASCII value of decimal 65
"1"        ' ASCII value of decimal 49
```

String Constants

Although PicBasic does not provide string-handling functions, we can define strings of characters by enclosing them in double quotes. For example,

```
"COMPUTER"
```

The above string is treated as a string of ASCII characters with values "C", "O", "M", "P", "U", "T", "E", "R".

Line Labels

In PicBasic programs, we often want to jump to different parts of a program, or to jump to a subroutine. A line in PicBasic is referred by a line label. A line label can be a valid identifier (a valid name in PicBasic), followed by a colon character (:). For example,

```
LOOP:
```

Multi-statement Lines

It is possible to use more than one statement on a line to make the program more readable. A colon (:) character should be used to separate more than one statement in a line. The size of the code does not change when more than one statement is written on the same line. For example, consider the following statements:

B0 = 3
B1 = 5
B2 = 8

The above statements can all be written on the same line as

B0 = 3   :   B1 = 5   :   B2 = 8

4.1.2   PicBasic mathematical and logical operations

PicBasic supports a number of mathematical and logical functions which make calculations easy in programs. The operations are performed on integer numbers only with 16-bit precision and there is no floating-point number format. Also, all math operations are performed strictly from left to right. The operators supported are

+	addition
−	subtraction
*	multiplication
**	most significant bit (MSB) of multiplication
/	division
//	remainder in a division
MIN	limit to minimum value
MAX	limit to maximum value
&	bitwise AND
|	bitwise OR
^	bitwise XOR
&/	bitwise AND NOT
|/	bitwise OR NOT
^/	bitwise XOR NOT

Multiplication is done on 16×16 bit numbers, resulting in a 32-bit result. The "*" operator returns the lower 16-bits of the 32-bit result. Similarly, the "**" operator returns the upper 16-bits of the result. For example,

W2 = W1 * W0   ' Multiply W1 with W0. The lower 16-bits of the result
                    ' are placed in W2

or,

W2 = W1 ** W0   ' Multiply W1 with W0. The upper 16-bits of the result
                    ' are placed in W2

or,

W2 = W1 * 100   ' Multiply W1 with 100. Place the lower 16-bits of the
                    ' result in W2. Note that this is the multiplication
                    ' found in most programming languages

Similarly with division,

W2 = W1 / W0   ' Divide W1 by W0. The result is placed in W2

or,

W2 = W1 // W0   ' Divide W1 by W0. The remainder is placed in W2

MIN is used to limit the result to the minimum value defined. For example,

B1 = B0 MIN 100

Sets B1 to the smaller of B0 and 100, i.e. B1 cannot be greater than 100.

Similarly, MAX is used to limit the result to the maximum value defined. For example,

B1 = B0 MAX 100

sets B1 to the larger of B0 and 100, i.e. B1 will be between 100 and 255.

Bitwise logical operations operate on the entire byte and these operations can be used to extract bits from bytes or to set and clear bits of a byte. For example, to extract the least significant bit of B0 we can write

B0 = B0 & %00000001

Similarly, to set bit 2 of B1 to be 1 we can write

B1 = B1 | %00000100

To store the upper four bits of B2 in B1 we can write

B1 = B2 & %11110000

4.1.3   PicBasic program flow control commands

Program flow control commands are important in every programming language since they enable the programmer to make a decision and change the flow of the program based on this decision. PicBasic language supports the following program flow control commands:

BRANCH
BUTTON
CALL
FOR…NEXT
GOSUB…RETURN
GOTO
IF…THEN

We shall now see what the functions of these commands are and how to use them in programs.

BRANCH

BRANCH *offset, (Label0, Label1,...)*

When this command is executed, the program will jump to the program *label* based on the value of *offset*. *Offset* is actually a program value and if *offset* is zero, the program jumps to the first label, if *offset* is one, the program jumps to the second label, and so on.

Example:

BRANCH B2, (Lbl1, Lbl2, Lbl3)   ' If B2 = 0 then goto Lbl1
                                  ' If B2 = 1 then goto Lbl2
                                  ' If B2 = 2 then goto Lbl3

BUTTON

BUTTON *Pin, Down, Delay, Rate, Var, Action, Label*

This command is used to check the status of a switch. The command operates in a loop and continuously samples the pin, debouncing it and comparing the number of iteration performed with the switch closed. The parameters are

Pin	Pin number (0 to 7). PORTB pins only
Down	State of pin when button is pressed (0 or 1)
Delay	Delay before auto-repeat begins (0 to 255). If 0, no debounce or auto-repeat is performed. If 255, only debounce, but no auto-repeat is performed
Rate	Auto-repeat rate (0 to 255)
Var	Byte variable used for delay/repeat countdown. Should be initialised to 0 before use
Action	State of pin to perform goto (0 if not pressed, 1 if pressed)
Label	Program execution continues at this label if *Action* is true

Figure 4.1 shows the two types of switches that can be used with this command.

For example, the following command checks for a switch pressed on pin 2 (of PORTB) and jumps to Loop if it is not pressed (this command assumes that the port pin will be logic 0 when the switch is pressed, i.e. the figure on the left in Figure 4.1):

BUTTON 2, 0, 255, 0, B0, 0, Loop

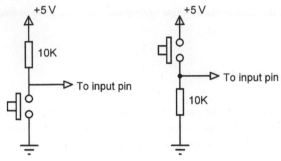

Figure 4.1   Switches that can be used for the Button command

The following command checks for a switch pressed on pin 2 as above, but jumps to Loop if the switch is pressed:

BUTTON 2, 0, 255, 0, B0, 1, Loop

CALL

CALL *Label*

This command executes the assembly language subroutine named *Label*. For example, the command calls to assembly language routine with the name *calculate*.

CALL calculate

FOR...NEXT

FOR *index* = *Start* TO *End* (STEP (−) *Inc*)
        (body)
NEXT *index*

This command is used to perform iterations in a program. *Index* is a program variable which holds the initial value of the iteration count *Start*. *End* is the final value of the iteration count. *STEP* is the value by which the *index* is incremented at each iteration. If no *STEP* is specified, the *index* is incremented by 1. The iteration repeats until *index* = *End* and then execution continues with the next instruction following the *NEXT*. *Index* can be a byte (0 to 255), or a word (0 to 65535).

In the following example, the two statements enclosed within the FOR...NEXT are executed 10 times.

FOR   B0 = 1 TO 10
        B1 = B1 + 1
        B2 = B2 + 1
NEXT B0

or in the following example, the *index* is incremented by 2 in each iteration.

```
FOR B0 = 1 TO 100 STEP 2
    B1 = B0 + 2
NEXT B0
```

GOSUB...RETURN

```
GOSUB Label
```

This program calls a subroutine starting at *Label*. It is like a GOTO command, but here the program returns when the RETURN statement is reached, and continues with the instruction after the GOSUB. The RETURN statement has no parameters. A subroutine has the following characteristics:

- A label to identify the starting point of the subroutine
- Body of the subroutine where the required operation is performed
- RETURN statement to exit the subroutine and return to the main calling program

Subroutines can be nested in PicBasic where a subroutine can call to other subroutines. The nesting should be restricted to no more than four levels deep. In the following example, the subroutine labelled *INC* increments variable B1 by one and then returns to the main program. On return to the main program, the statement B2 = B1 is executed.

```
        B0 = 0
        B1 = 1
        GOSUB INC      ' Jump to subroutine INC
        B2 = B1        ' Subroutine returns here
        .........
        .........
INC:                   ' Start of the subroutine
        B1 = B1 + 1    ' Body of the subroutine
        RETURN         ' End of the subroutine
```

GOTO

```
GOTO Label
```

This command causes the program execution to jump to the statement beginning at *Label*. For example,

```
        GOTO Loop
        .........
        .........
        .........
Loop:
```

IF...THEN

```
IF Comp (AND / OR Comp) THEN Label
```

This statement is used to perform comparisons (*Comp*) in a program. If the result of the comparison is true then the program jumps to the statement at Label, otherwise execution resumes with the statement following IF...THEN.

A comparison can relate to a variable, to a constant, or to other variables. All comparisons are unsigned and the following comparison operators can be used:

<	less than
<=	less than or equal
=	equal
<>	not equal
>=	greater than or equal
>	greater than

Additionally, logical operators AND and OR can be used in a comparison operation. For example,

        IF B0 > 10 THEN CALC   ' Jump to CALC if B0 > 10
        ....
        ....

CALC:

Another example is given below. In this example, if B2 is greater than 40 and at the same time B3 is less than 20 then the program jumps to the statement at label EXT. Otherwise, execution continues with the statement after the IF...THEN.

        IF B2 > 40 AND B3 < 20 THEN EXT
        ....
        ....

EXT:

It is important to be careful that only a *Label* can be used after the THEN statement.

4.1.4   Other PicBasic commands

We shall now briefly look at the remaining PicBasic commands in alphabetical order which are useful during the program development. More details about these commands can be obtained from the PicBasic manual.

EEPROM

        EEPROM *Location, (constant, constant,....., constant)*

This command stores constants in consecutive bytes in on-chip EEPROM memory. The command only works with the PIC microcontrollers that have EEPROM, such as the PIC16F84, PIC16F877, etc. Location is optional, and if omitted the first EEPROM location is assumed. Constants can be numeric constants or string constants. Strings are stored as consecutive bytes of ASCII values. An example is given below.

> EEPROM 3, (5, 2, 8)   ' Store 5 in location 3,
> '2 in location 4, and 8 in
> 'location 5

END

> END

Stops execution and enters low power mode. The command has no parameters.

HIGH

> HIGH *Pin*

Makes the specified pin an output pin and sets it to logic 1. *Pin* only applies to PORTB pins and it can take values from 0 to 7. In the following example, bit 1 of PORTB is configured as an output pin and is set to logic 1:

> HIGH 1

I2CIN

> I2CIN *Control, Address, Var, (,Var)*

This command is used to read data from serial EEPROMs with a 2-wire I^2C interface. A list of some compatible devices is given in Table 4.4. The lower 7 bits of the *Control* byte contain a 4-bit control code, followed by the chip select or additional address information, depending on the device used. As shown in Table 4.4, the 4-bit control code for EEPROMs is "1010". The high-order bit (MSB) of the *Control* byte is a flag indicating whether the *Address* is to be sent as 8 bits or 16 bits. If the flag is low, the *Address* is sent as 8 bits, and if it is high, the *Address* is sent as 16 bits. *(,Var)* shown in the command list is used only for 1-bit information. The I^2C data and clock lines are predefined in the PicBasic library as bit 0 of PORTA (RA0) and bit 1 of PORTA (RA1), respectively.

For example, when communicating with a 24LC02B EEPROM, the required *Address* is 8 bits, the control code is "1010" and chip select or additional address information is not required and can be assumed to be 0. The required *Control* byte is then "01010000".

Figure 4.2 shows how the 24LC02B (or any other serial EEPROM) can be connected to a PIC microcontroller. In this example, a PIC16F84 is used and pin RA0 and RA1 are connected to the data and clock pins of the EEPROM, respectively. These are the only connections required to communicate with an I^2C-compatible device. As shown in the figure, the I^2C lines should be connected to Vdd (+5 supply) with 4.7K resistors.

Table 4.4   Some I²C compatible EEPROMs

Device	Capacity	Control	Address size
24LC01B	128 bytes	01010xxx	8 bits
24LC02B	256 bytes	01010xxx	8 bits
24LC04B	512 bytes	01010xxb	8 bits
24LC08B	1 K bytes	01010xbb	8 bits
24LC16B	2 K bytes	01010bbb	8 bits
24LC32B	4 K bytes	11010ddd	16 bits
24LC65B	8 K bytes	11010ddd	16 bits

bbb = block select bits (each block is 256 bytes)
ddd = device select bits
xxx = don't care

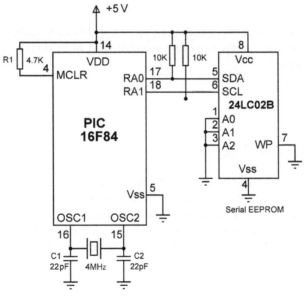

Figure 4.2   I²C Connections to a PIC microcontroller

In the following example, a data byte is read from address 20 of the serial EEPROM and stored in variable B1. Note that the *Control* byte is set to "01010000", *Address* is assigned variable B0 and value 20 stored in it, and the byte read from the EEPROM is stored in data register B1.

```
Symbol con = %01010000
Symbol addr = B0
        addr = 20            ' Set address to 20
        I2CIN con, addr, B1  ' Read from address 20 to B1
```

I2COUT

> I2COUT *Control, Address, Value (,Value)*

This command is used to send data to an I²C compatible device such as a serial EEPROM described in command I2CIN. The *(,Value)* in the command is used for 16-bit information.

When writing data to an EEPROM, it is necessary to wait about 10 ms (device dependent) for the write operation to complete before attempting to write again. In the example given below, data byte 10 is written to address 30, and also data byte in variable B5 is written to address 31 of an EEPROM.

```
Symbol con = %01010000
Symbol addr = B0

          addr = 30                 ' Set address to 30
          I2COUT con, addr, (10)    ' Write byte 10 to address 30
          PAUSE 10                  ' Wait 10ms

          addr = 31                 ' Set address to 31
          I2COUT con, addr, (B5)    ' Write byte in B5 to address 31
          PAUSE 10                  ' Wait 10 ms
```

INPUT

> INPUT *Pin*

This makes the specified PORTB pin an input. *Pin* is from 0 to 7. For example,

> INPUT 2   ' Make RB2 an input pin

LOOKDOWN

> LOOKDOWN *Search, (Constant, Constant,......), Var*

This command provides a look-up table. It looks down a list of *Constants* and compares each one with the *Search* value. If a match is found, the position of the match is stored in *Var*. Note that the first *Constant* is assumed to be at position 0. The *Constant* list can be numeric or string constants. In the following example, if we assume that variable B0 has value 5 then variable B1 will contain 3 which is the position of 5 in the table:

> LOOKDOWN B0, (0, 8, 9, 5, 12, 0, 1), B1

LOOKUP

> LOOKUP *Index, (Constant, Constant,.....), Var*

This command is used to retrieve values from a table. When *Index* is 0, *Var* is loaded with the first *Constant*; when *Index* is 1, *Var* is loaded with the second *Constant* and so on. In the following example, if we assume that variable B0 has value 3, variable B1 will be loaded with 8 which is the 3rd element in the table starting from 0:

LOOKUP B0, (0, 9, 0, 8, 12, 32), B1

LOW

LOW *Pin*

This command makes the specified pin an output pin and clears it to logic 0. *Pin* only applies to PORTB pins and it can take values from 0 to 7. In the following example, bit 2 of PORTB is configured as an output pin and is cleared to logic 0:

LOW 2

NAP

NAP *Period*

The NAP command places the PIC microcontroller in low-power mode for a while to save power in battery applications. The Period is a variable from 0 to 7 and the approximate delay is given in Table 4.5.

Table 4.5   Delay in NAP command

Period	Delay (s, approx)
0	18×10^{-3}
1	36×10^{-3}
2	72×10^{-3}
3	144×10^{-3}
4	288×10^{-3}
5	576×10^{-3}
6	1.152
7	2.304

In the following example the microcontroller is put into low power mode for just over 1 s:

NAP 6

OUTPUT

OUTPUT Pin

This command makes the specified pin of PORTB an output pin. Pin can take values from 0 to 7. In the following example, bit 2 of PORTB (RB2) is made an output pin:

OUTPUT 2

PAUSE

PAUSE Period

This is one of the commonly used commands to delay a program by a specified amount. Period is in milliseconds and can range from 1 to 65,535 ms (i.e. just over one minute). PAUSE does not put the microcontroller into low-power mode. In the following example, the program is delayed by 1 s:

PAUSE 1000

PEEK

PEEK *Address, Var*

This command is used to read the value of a RAM register at the specified Address and then put the value into variable Var. The PEEK command can be used to access all registers of the PIC microcontroller including the Port registers, A/D converter registers, etc.

In the following example, the 8-bit value of PORTB is read and stored in variable B0:

Symbol PORTB = 6   ' PORTB register address
PEEK PORTB, B0   ' Read PORTB into B0

POKE

POKE *Address, Var*

This command is used to send data to a RAM register at the specified Address. The POKE command can be used to send data to all accessible registers of the PIC microcontroller, including the PORT registers, PORT direction registers, A/D converter registers, etc.

In the following example, TRISB is cleared to 0 so that all PORTB pins are outputs. The hexadecimal value 24 is sent to PORTB.

Symbol TRISB = $86   ' TRISB register address
Smbol PORTB = 6   ' PORTB register address

POKE TRISB, 0   ' Clear TRISB
POKE PORTB, $24   ' Send $24 to PORTB

POT

POT *Pin, Scale, Var*

This command could be useful to read an analogue voltage if the microcontroller has no built-in A/D converter. *Pin* is a PORTB pin and can take a value between 0 and 7. For this command to work, a resistor and a capacitor are serially connected to a port pin as shown in Figure 4.3. When a voltage is applied to a resistor–capacitor circuit, the voltage across the capacitor rises exponentially as the capacitor is charged through the resistor. The charge time is dependent on the value of the resistor and the capacitor.

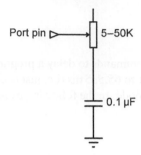

Figure 4.3   Resistor and capacitor connected to an I/O pin

When the POT command is used, the capacitor is initially discharged by the I/O pin by placing the pin in output mode. After that, the I/O port is changed to an input port and starts timing the voltage across the capacitor until the voltage reaches the threshold value of the I/O pin. When this happens, the calculated charge time is converted into a number between 0 and 255 and is stored in *Var*. The *Scale* value should be set experimentally. To do this, set the device to maximum resistance and set the *Scale* to 255. The value returned in *Var* will be the proper scale value for the chosen components. An example is given below where the resistor–capacitor are connected to pin 1 of PORTB, the *Scale* value is set to 255 and the output value is stored in B0.

> POT 1, 255, B0

PULSIN

> PULSIN *Pin, State, Var*

The PULSIN command measures the pulse width of any signal connected to a PORTB pin. With a 4 MHz crystal or resonator, the pulse width will be measured in 10 μs units. If *State* is 0, the width of a low pulse is measured; if *Scale* is 1, the width if a high pulse is measured. The measured value in 10 μs units is stored in variable *Var*. *Var* can be a byte or a word. If a word is used, it can take values 1 to 65,535, i.e. the minimum pulse width that can be measured is 10 μs and the maximum is 655,350 μs. If a byte is used, the range of the measurement is 10 to 2550 μs.

PULSOUT

> PULSOUT *Pin, Period*

This command generates a pulse on a PORTB pin (*Pin* can be 0 to 7) of specified *Period* in 10 μs units. The *Period* is a word and thus pulses of up to 655,350 μs can be generated. The specified pin is automatically made an output pin.

For example, to generate a 500-μs pulse on pin 1 of PORTB, we need the command

> PULSOUT 1, 50

PWM

> PWM Pin, Duty, Cycle

This command outputs a Pulse-Width-Modulated (PWM) signal on the specified PORTB pin (*Pin* can be 0 to 7). The *Duty* is the pulse duty-cycle and can range from 0 to 255. 0 corresponds to a 0% duty-cycle, and 255 corresponds to a 100% duty-cycle. The generated PWM pulse is repeated *Cycle* times. The specified port *Pin* is made an output just before the command is executed and reverts to an input after the pulse is generated.

In the following example, a 200-cycle PWM signal is generated on bit 0 of PORTB with a duty-cycle of 50%:

PWM 0, 127, 200

Another use of this command is to generate an analogue signal by sending the output to a resistor–capacitor circuit as shown in Figure 4.4. In this circuit, the voltage across the capacitor will vary depending on the *Duty* and the *Cycle* of the pulses.

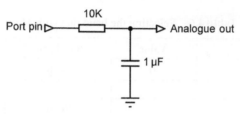

Figure 4.4 Using PWM signal for D/A conversion

RANDOM

RANDOM *Var*

This command generates a random number and stores in word variable *Var*. For example, to generate a random number and store in W1 use the command:

RANDOM W1

READ

READ *Address, Var*

This command is used to read a byte from the specified *Address* of the built-in EEPROM memory. The byte read is stored in variable *Var*. This command can only be used with PIC microcontrollers that have built-in EEPROM memory (such as PIC16F84, or PIC16F877).

In the following example, the byte at address 10 of EEPROM is read and stored in variable B1:

READ 10, B1   'Read byte at address 10
                   'and store in B1

REVERSE

REVERSE *Pin*

This command reverses the mode of a PORTB pin (*Pin* can be from 0 to 7). If the pin is an input, it is made an output. Similarly, if the pin is an output, it is made an input.

In the following example, bit 2 of PORTB is first made an output pin, then changed to an input pin:

OUTPUT 2   'RB2 is output pin
REVERSE 2   'RB" is an input pin

SERIN

SERIN *Pin, Mode, (Qual, Qual,,), Item, Item,*

This command is used to receive RS232 serial asynchronous data on a PORTB pin (pin is between 0 and 7) using 8-bit data, no parity bit, and one stop bit. As shown in Table 4.6, *Mode* defines the baud rate and whether or not the pin data is inverted. For example, if Mode is N9600, the data is inverted and the selected baud rate is 9600.

Table 4.6   Selecting the baud rate with *Mode*

Symbol	Value	Baud rate	Mode
T2400	0	2400	True
T1200	1	1200	True
T9600	2	9600	True
T300	3	300	True
N2400	4	2400	Inverted
N1200	5	1200	Inverted
N9600	6	9600	Inverted
N300	7	300	Inverted

The RS232 signal levels are ± 12 V and level converter circuits (such as MAX232) are normally used to convert the RS232 signal levels to TTL and the TTL levels back to RS232 levels. The I/O specifications of PIC microcontrollers allow RS232 signals to be directly connected to a port pin. As shown in Figure 4.5, a resistor is all that is needed to receive RS232-compatible signals on a pin. When used in this mode, the data is to be inverted (i.e. use the "N" versions of the mode signals in Table 5.6)

Figure 4.5   Connecting a RS232 signal to a port pin

A number of qualifiers, enclosed in brackets, can be used with the SERIN command such that these bytes must be received before receiving the data items. Once the qualifiers are satisfied, SERIN receives the serial data and stores in *Item*s. The *Item* variable may be preceded by the hash character ("#"). This will convert the decimal number received into ASCII equivalent and store it in *Item*.

In the following example, pin 1 of PORTB (RB1) is defined as the serial I/O pin and the port pin is connected to the RS232 serial line using a resistor. The baud rate is assumed to be 4800. The microcontroller waits until the character "X" is received from the line and then stores the next byte in variable B0:

SERIN 1, N4800, ("X"), B0

SEROUT

SEROUT *Pin, Mode, (Item, Item,...)*

This command is similar to the SERIN command but is used to send RS232 asynchronous serial data to a pin of PORTB (*Pin* can be between 0 and 7). As before, *Mode* is used to set the communications baud rate. In addition to the standard inverted and non-inverted modes, it is also possible to set Open-Drain and Open-Collector modes where a pull-up resistor will be required at the output of the pin. Table 4.7 gives a list of the available *Modes*.

Table 4.7 Selecting the baud rate with *Mode*

Symbol	Value	Baud rate	Mode
T2400	0	2400	True
T1200	1	1200	True
T9600	2	9600	True
T300	3	300	True
N2400	4	2400	Inverted
N1200	5	1200	Inverted
N9600	6	9600	Inverted
N300	7	300	Inverted
OT2400	8	2400	Open Drain
OT1200	9	1200	Open Drain
OT9600	10	9600	Open Drain
OT300	11	300	Open Drain
ON2400	12	2400	Open Source
ON1200	13	1200	Open Source
ON9600	14	9600	Open Source
ON300	15	300	Open Source

Data byte Item is sent to the specified port pin in serial format. The Item can be a string constant or a numeric value. A string constant consists of characters and each character of the string is sent out. For example, the string "COMPUTER" is sent out as 8 individual characters. A numeric value will send the corresponding ASCII character. For example, 13 is the carriage-return character, 65 is character "A" and so on. A numeric value can be preceded by the hash character "#" and this will send out the ASCII representation of its decimal value. For example, #345 will be sent as "3", "4", and "5".

In the following example, it is assumed that pin 1 of PORTB (RB1) is used as the serial I/O pin and it is configured for 4800 baud. ASCII value of variable B0 is sent out from this pin, followed by a carriage-return.

    SEROUT 1, N4800, (#B0, 13)

SLEEP

    SLEEP *Period*

The SLEEP command is used to put the microcontroller in low-power mode and stops the microcontroller running for the specified *Period*. The *Period* is a word and can range from 1 to 65,535 and represents increments of 2.3 s. For example, a value of 1 will make the microcontroller sleep for 2.3 s, a value of 2 will make the microcontroller sleep for 4.6 s and so on. The maximum value of 65,535 makes the microcontroller sleep just over 18 h.

In the following example, the microcontroller sleeps for 23 s:

    SLEEP 10

SOUND

    SOUND *Pin, (Note, Duration, Note, Duration,.....)*

This command is used to generate sound on a specified PORTB pin of the microcontroller (*Pins* are between 0 and 7). Note can take values from 0 to 255 and these values do not correspond to the musical notes. A 0 represents silence. Values from 1 to 127 are tones (1 is lower frequency than 127), and values from 128 to 255 are white noise (128 is lower frequency than 255). The sound continues for a length of time specified by Duration. Duration is measured in milliseconds and it can take values between 0 and 255. The SOUND command produces TTL level square waves and it is possible to connect a speaker to the output pin as shown in Figure 4.6.

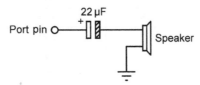

Figure 4.6   Connecting a speaker for the SOUND command

In the following example, a sound with note 20 and duration 100 ms is sent to pin 0 of PORTB. Then, another sound with note 23 and duration 200 ms is sent out from the same port pin.

    SOUND 0, (20, 100, 23, 200)

TOGGLE

    TOGGLE *Pin*

This command makes the specified *Pin* an output pin and inverts the state of this pin (*Pin* can take values from 0 to 7).

In the following example, bit 0 of PORTB (RB0) is first made low, and then changed to high using the TOGGLE command:

```
LOW 0
TOGGLE 0
```

WRITE

WRITE *Address, Value*

The WRITE command writes the Value byte to the specified EEPROM address. This command is only valid for the PIC microcontrollers which have built-in EEPROM memories.

In the following example byte in variable B0 is written to EEPROM address 2:

```
WRITE 2, B0
```

4.1.5   Recommended PicBasic program structure

There are many different ways in which a PicBasic program can be written. It is important to note that a program should be written in such a way that it is easily maintainable by other people. This is specially important if you work in a firm and others may have to upgrade or maintain your program. The following steps should be followed to develop a maintainable program:

- Use a header in your programs. This header should briefly describe the function of the program. In addition, the author of the program, the program creation date, program file name, and any program modifications should be described in the header.
- Use comments in your programs to describe what you are trying to do. The comments can be used at the beginning of a piece of code, or after every statement.
- Use symbols as much as possible in your programs. Symbols make your programs more readable.

The author recommends that you use a template similar to the one given in Figure 4.7 when developing PicBasic programs. As you can see in this figure, the header includes a brief description of the program, name of the author, the date, and the filename of the program. Comments are used in every line of the program to clarify the actions of the program.

4.2   PicBasic Pro language

PicBasic Pro is a full-featured compiler and is for serious or professional PIC programmers. PicBasic Pro has many additional commands compared to the standard PicBasic compiler. In addition, the variables, constants and symbols are treated differently in PicBasic Pro. In this section, we shall only be looking at the commands which are specific to PicBasic Pro language, and which have not been described in Section 4.1. Also, various features of the PicBasic Pro language are described in this section.

```
' ************************************************************************
'
'                          LED FLASHING PROGRAM
'                          ========================
'
'
' This program flashes and LED connected to port RB0 of PORTB. The
' Led is flashed with 1 second intervals.
'
' Author:        Dogan Ibrahim
' Date:          September, 2005
' File:          LED.PBC

' Modifications
' ===========
'
'
' ************************************************************************
'
' SYMBOLS
'
Symbol LED = 0                    ' Define RB0 as LED
Symbol TRISB = $86               ' TRISB address
Symbol PORTB = $06               ' PORTB address
'
' START OF MAIN PROGRAM
'
        POKE TRISB, 0            ' Set PORTB pins as outputs
AGAIN:
        HIGH LED                ' Turn ON LED
        PAUSE 1000              ' Wait 1 second

        LOW LED                 ' Turn OFF LED
        PAUSE 1000              ' Wait 1 second

        GOTO AGAIN              ' Repeat

        END                     ' End of program
```

Figure 4.7   Recommended PicBasic program template

4.2.1   PicBasic Pro variables

Variables in PicBasic Pro are stored in the general purpose RAM registers and are declared using the VAR keyword. Each variable has a name and a variable type. A variable type can be a bit, a byte, or a word. Some example variable declarations are

```
Total     VAR     word
Count     VAR     byte
Flag      VAR     bit
```

The VAR keyword can also be used to create an alias for a variable (i.e. another name). In the following example, *Sum* is another name for *Total*:

    Sum   VAR   Total

The individual bits of a variable can be accessed by writing the variable name, followed by a dot "." character, and then the bit number (0 to 15), or the keyword BIT followed by the bit number (e.g. BIT0 to BIT15). The following are examples of accessing bit 0 of variable *Total*:

    Total.0
    Total.BIT0

Arrays of variables can be created in PicBasic Pro by writing the name of the array, followed by the keyword VAR, and then the type and the size of the array. For example, a byte array called *Sum* with 10 elements of type *byte* can be declared as

    Sum   VAR   byte[10]

In the above example, the first element of the array is *Sum[0]*, and the last element is *Sum[9]*. Arrays have a size-limit in PicBasic Pro.

- Maximum size of a bit array is 256
- Maximum size of a byte array is 96 (microcontroller-dependent)
- Maximum size of a word array is 48 (microcontroller-dependent)

4.2.2 Constants

Constants in the PicBasic Pro language are declared using the CON keyword. A constant value cannot be changed in a program.

In the following example, *Maxim* is declared as 10 and its value cannot be changed in the program:

    Maxim   CON   10

4.2.3 Comments

Comments in PicBasic Pro are declared same as in the PicBasic language, i.e. using the REM keyword or a single quote at the beginning of a line.

4.2.4 Multi-statement lines

Multi-statement lines are created as in PicBasic, i.e. by separating each statement with a colon ":" character.

4.2.5 INCLUDE

Other PicBasic Pro source files can be included in a program as in PicBasic language.

4.2.6 DEFINE

This command defines various compiler options, such as the clock oscillator frequency, pin number, etc.

4.2.7 Line extension

When writing long programs, it may be necessary to continue part of a statement on a new line. A line can be extended by typing the line extension character "_" as the last character in the line to be continued. For example,

```
Item1, Item2, Item3, Item4, _
Item 5, Item6
```

4.2.8 Accessing ports and other registers in PicBasic Pro

PIC microcontroller ports or any other registers can easily be accessed by simply writing the name of the port or the register and using equate "=" character. For example,

```
A = PORTA
UP = PORTB & $F0
PORTB = $2F
INTCON = $0F
```

The bits of a port or a register can be accessed by simply writing the name of the port or the register, followed by a dot "." character and the PORTBit to be accessed. For example,

```
L = PORTB.1        ' Read bit 1 of PORTB and load into L
L = PORTB.BIT1     ' Read bit 1 of PORTB and load into L
K = STATUS.0       ' Read bit 0 of STATUS register and load into K
```

In most of the PicBasic I/O commands, *Pin* is used to define a pin of PORTB where *Pin* can take a value from 0 to 7 corresponding to PORTB pins. In a similar manner, PicBasic allows the use of numbers 0 to 15 to access port I/O pins. When only a number is used to access a port pin, the port and the pin number accessed depends on the package size of the microcontroller used. Table 4.8 shows the Pin definitions for 8 to 40 pin PIC microcontrollers.

For example, assuming we are using an 18-pin PIC microcontroller, the PicBasic command

```
SOUND 3, 10,100
```

generates a sound with note 10 and duration 100 ms from bit 3 of PORTB (i.e. RB3). In PicBasic Pro, we can use the same statement, or we can write

    SOUND PORTB.3, 10, 100

If we wish to generate the sound from bit 0 of PORTA, in PicBasic Pro we can write

    SOUND PORTA.0, 10, 100

or

    SOUND 8, 10, 100

There is no way of generating a sound from PORTA using the PicBasic language.

Table 4.8   Port I/O *Pin* definitions

PIC micro Size	Pin 0–7	Pin 8–15
8 pin	GPIO	–
18 pin	PORTB	PORTA
24 pin (except 14,000)	PORTB	PORTC
28 pin	PORTC	PORTD
40 pin	PORTB	PORTC

The direction of a port is determined by loading the corresponding TRIS register. For an output pin, a 0 is loaded into the corresponding TRIS register, and for an input pin a 1 is loaded into the corresponding TRIS register. In PicBasic Pro, the TRIS register can be accessed directly like any other register. For example, to configure all PORTB pins as outputs and then send the hexadecimal value $FF to PORTB we can write

    TRISB = 0
    PORTB = $FF

4.2.9   Arithmetic operators

PicBasic Pro supports more arithmetic operators than PicBasic. Table 4.9 lists all the arithmetic operators supported by PicBasic Pro. In this section, we shall be looking only at these additional operators.

Shift

The shift operators "<<" and ">>" are used to shift a value left or right, respectively, 0 to 15 times. Zeroes are placed to the shifted positions. Shifting left is same as multiplying the number by 2, and shifting right is same as dividing the number by 2.

Table 4.9   PicBasic Pro arithmetic operators

Arithmetic Operator	Description
+ − * /	Add, subtract, multiply, divide
**	Top 16 bits of multiplication
*/	Middle 16 bits of multiplication
//	Remainder
<< >>	Shift left, shift right
ABS	Absolute value
COS	Cosine
DCD	Decode
DIG	Digit
MAX MIN	Maximum, minimum
NCD	Encode
REV	Reverse bits
SIN	Sine
SQR	Square root
& | ^ ~	Bitwise AND, OR, EXOR, NOT
&/ |/ ^/	Bitwise NAND, NOR, INOR

In the following first example, variable *Cnt* is shifted left twice. In the second example, variable *Sum* is shifted right 3 times.

```
Cnt = cnt << 2      ' Shift left Cnt by 2 places
Sum = Sum >> 3      ' Shift right Sum by 3 places
```

ABS

Operator ABS returns the absolute value of a number. In the following example, the absolute value of variable *p* is returned:

```
p = ABS p           ' Return the absolute value
```

COS

Returns the cosine of a number. The result is in 2's complement format in the range −127 to +127. The number must be in radians in the range 0 to 255. In the following example, the cosine of 8 radians is returned:

```
Angle = COS 8       ' Return the cosine of 8
```

DCD

This operator is used to set a bit of a byte or a word to 1. All other bits are set to 0. For example, to set bit 4 of a byte we can write

      B4 = DCD 4   ' Set bit 4 of variable B4

Where variable B4 will take the binary value %00010000

DIG

This operator returns a digit of a number. The number can be up to 4 digits with the rightmost digit being digit 0. For example, if variable *Sum* is equal to 678, the first digit (number 7) can be extracted as

      Sum = 678   ' Sum = 678
      P = Sum DIG 1   ' P = 7

NCD

The NCD operator is used to find the highest bit number set in a number. The bit numbers can range from 1 to 16. A zero is returned if no bit is set. In the following example, variable P = 6 since the highest bit set in the number is the sixth bit (starting from 1).

      P = NCD %00101011   ' Highest bit set is 6

SIN

This operator is similar to the COS operator and it returns the sine of a number. The number must be expressed in radians and it must be between 0 and 255. For example, to find the sine of 10 radians, use

      P = SIN 10

SQR

This operator returns the square root of a number. The result is an integer number. For example, to find the square root of variable *Total*, use

      N = SQR Total   ' Find square root of *Total*

4.2.10   *PicBasic Pro commands*

PicBasic Pro has over 80 commands. Some commands are similar to the PicBasic commands with minor changes. For example, the range of the Pin variable is from 0 to 15, instead of 0 to 7. It is the author's recommendation that you use the port name, followed by a dot and the bit number when you wish to access a port pin. This makes your programs much more readable and easier to maintain.

In this section, we shall only look at the commonly used commands which are specific to the PicBasic Pro language. Further information about these or any other commands can be obtained from the PicBasic Pro user manual.

ADCIN

ADCIN *Channel, Var*

This command is used to read the on-chip A/D converter. This is not a very useful command and we shall see in the projects section how to read data from the A/D channel of a PIC microcontroller.

BRANCHL

BRANCHL *Index, (Label, Label,......)*

The BRANCH command used in the PicBasic language causes a limited range of branch (usually 1 K). The BRANCHL command can be used to create longer jumps in the program memory. The BRANCHL command is slower than the BRANCH command and generates more assembly code.

CLEAR

CLEAR

This command clears (zeroes) all the RAM registers in each bank.

CLEARWDT

CLEARWDT

If the watchdog timer is enabled, it can time out and reset the program to the beginning (address 0). The CLEARWDT command is used to reset the watchdog timer so that it does not time out.

COUNT

COUNT *Pin, Period, Var*

This command is used to count the number of pulses that occur on *Pin* during the *Period* and stores the result in *Var*. *Pin* can take values 0 to 15 but the "*Portname.number*" format is recommended (e.g. PORTB.0).

The highest frequency that can be counted with a 4 MHz crystal clock is 25 kHz, and 125 kHz when a 20 MHz clock is used. In the following example, the number of pulses on bit 0 of PORTB are counted in 100 ms and stored in variable *Cnt*:

COUNT PORTB.0, 100, Cnt

DATA

DATA *@Location, Constant, Constant,....*

This command stores constants in the on-chip EEPROM memory during the programming of the device (not when the program is run). The command can only be used with the PIC microcontrollers that have on-chip EEPROMs. *Location* denotes the starting address of the EEPROM and if omitted, address 0 is assumed.

The following example shows how the numbers 5, 10, 15, and 20 can be stored in EEPROM starting from address 6:

DATA @6, 5, 10, 15, 20

DTMFOUT

DTMFOUT *Pin, Onms, Offms, [Tone, Tone,.....]*

This command produces Touch Tones normally available in keyboards and mobile phones. *Pin* can take a value between 0 and 15 (or *Portname.number*) and the specified pin is made an output. *Onms* is the duration of each tone in milliseconds, and *Offms* is the number of milliseconds pause between each tone. If the *Onms* or the *Offms* are not specified, they default to 200 ms and 50 ms, respectively.

A *Tone* can take a value between 0 and 15. Tones 0 to 9 are the same as on a telephone keypad. *Tone* 10 is the * key, *Tone* 11 is the # key, and *Tones* 12–15 are the extended keys A to D. The sound generated by the DTMFOUT should be smoothed using resistor–capacitor filters. It is recommended to use a high clock rate (e.g. 20 MHz) to get a smooth signal after the filtering.

In the following example, the DTMF tones for numbers 886 are sent from bit 0 of PORTB with the default duration and pause:

DTMFOUT PORTB.0, [8, 8, 6]

FREQOUT

FREQOUT *Pin, Onms, Frequency1, [,Frequency2]*

This command generates a signal with one or two different frequencies on the specified *Pin* for *Onms* milliseconds. *Pin* is automatically made an output and it can be 0 to 15 or a *Portname.number*. The generated signal is a square wave and filtering may be required to obtain a smooth signal.

In the following example, a 1 kHz signal is generated on port 0 of PORTB for 3 s:

FREQOUT PORTB.0, 3000, 1000

HPWM

HPWM *Channel, Dutycycle, Frequency*

Some PIC microcontrollers have one or more built-in circuits to generate pulse width–modulated square-wave signals (PWM). For example, PIC16F877 has two PWM *Channel*s. Channel 1 is known as CCP1 (also PORTC.2) and Channel 2 is known as CCP2 (also PORTC.1).

Dutycycle can vary from 0 to 255 which corresponds to 0% (low all the time) to 100% (high all the time), respectively. A value of 127 gives 50% duty cycle. The highest *Frequency* is 32,767 Hz, and on microcontrollers with two channels, the *Frequency* must be the same on both channels.

The PWM signal is output from the specified pin continuously in the background while the program executes other instructions.

In the following example, a 1 kHz, 50% duty cycle PWM signal is generated from Channel 1 (CCP1) of a PIC16F877 type microcontroller:

HPWM 1, 127, 1000

HSERIN
HSERIN2

These commands are only available on microcontrollers that have built-in serial port devices such as an USART. The use of these commands is complicated and more details can be obtained from the PicBasic Pro user manual.

HSEROUT
HSEROUT2

These commands are only used on microcontrollers that have built-in serial port devices such as an USART. The commands are used to send out serial asynchronous data from the microcontroller with the required format. The use of these commands is complicated and more details can be obtained from the PicBasic Pro user manual.

IF..THEN..ELSE

These commands are similar to the PicBasic IF..THEN command but the PicBasic Pro language provides more flexibility when one or more comparisons are made. These commands can be used in the following formats:

Format 1:
        IF Comparison [AND/OR Comparison…] THEN Label
Format 2:
        IF Comparison [AND/OR Comparison…] THEN Statement…..
Format 3:
        IF Comparison [AND/OR Comparison…] THEN
                Statement….
        ELSE
                Statement
        ENDIF

Some examples for the use of this command are given below:

Conditional statement:
        IF PORTB.0 = 0 THEN Led = 1

Conditional jump:
        IF (PORTB.0 = 0) AND (PORTB.1 = 1) THEN Loop

Multiple statements:
        IF Cnt < 10 THEN A = A + 1: B = B + 1

Multiple statements:
        IF SUM < 10 THEN
                Cnt = Cnt + 1
                Tot = Tot + 1
        ENDIF

IF..THEN..ELSE

```
IF Total = 100 THEN
        Flag = 1
ELSE
        Flag = 0
ENDIF
```

PAUSEUS

PAUSEUS *Period*

This command pauses the program for Period microseconds. Period is a word in the range 1 to 65,535. Thus, the maximum delay is 65.535 ms. PAUSEUS command assumes that we are using a 4 MHz clock. The minimum delay that can be generated with PAUSEUS using a 4 MHz clock is 24 μs.

REPEAT..UNTIL

```
REPEAT
        Statement...
UNTIL Condition
```

This command is used to create loops in programs. The statements between the REPEAT and UNTIL are executed until the specified *Condition* is true.

In the following example, the statements between REPEAT and UNTIL are executed 10 times:

```
k = 0
REPEAT
        Sum = Sum + 1
        Cnt = Sum
        k = k + 1
UNTIL k < 10
```

SELECT..CASE

```
SELECT CASE Var
        CASE Expr1 [,Expr...]
                Statement...
        CASE Expr2 [,Expr...]
                Statement...
        [CASE ELSE
                Statement...]
END SELECT
```

This command is used instead of using multiple IF..THEN commands. The variable *Var* is compared with different values (or ranges of values) and an action is taken based on its value. If *Var* does not match any of the conditions, then the statements after the CASE ELSE are executed. The IS keyword is used after CASE to specify a comparison other than equal to.

In the following example, if x is 1, B is set to 100. If x is 2, B is set to 6. If x is 3 or 4, B is set to 50. If x is greater than 120, B is set to 1. If x is none of these, then B is set to 0.

```
SELECT CASE x
     CASE 1
          B = 100
     CASE 2
          B = 6
     CASE 3, 4
          B = 50
     CASE IS > 120
          B = 1
     CASE ELSE
          B = 0
END SELECT
```

SHIFTIN

SHIFTIN *Datapin, Clockpin, Mode, [Var{\bits}, Var{\bits},…]*

The SHIFTIN command is used to read data one bit at a time as clock is sent out to the sending device. The received data is stored in variables *Var*. *Datapin* is either from 0 to 15 or a *Portname. number* and specifies the pin number which is to receive the data. *\bits* optionally specify the number of bits to shift in and if omitted, 8 bits are assumed. *Clockpin* is either 0 to 15 or a *portname. number* and specifies the pin number where the clock is sent out. *Mode* has a value between 0 and 7 and it specifies the mode of the clock operation as shown in Table 4.10. For *Mode*s between 0 and 3, the clock output is normally low and goes high to clock in a bit, then returns low. For *Mode*s between 4 and 7, the clock output is normally high and goes low to clock in a bit, then returns high.

Table 4.10   SHIFTIN command clock *Mode*s

Mode No.	Operation
0	Shift in MSB first. Read before sending clock. Clock normally low
1	Shift in LSB first. Read before sending clock. Clock normally low
2	Shift in MSB first. Read after sending clock. Clock normally low
3	Shift in LSB first. Read after sending clock. Clock normally low
4	Shift in MSB first. Read before sending clock. Clock normally high
5	Shift in LSB first. Read before sending clock. Clock normally high
6	Shift in MSB first. Read after sending clock. Clock normally high
7	Shift in LSB first. Read after sending clock. Clock normally high

In the following example, data bits are received into bit 0 of PORTB and stored, LSB first, followed by 8 data bits in variable B1. *Mode* 0 is used here and the clock is sent out from bit 1 of PORTB.

SHIFTIN PORTB.0, PORTB.1, 0, [B1\8]

SHIFTOUT

> SHIFTOUT *Datapin, Clockpin, Mode, [Var{\bits}, Var{\bits},....]*

This command is similar to SHIFTIN, but here, data bits are sent out one bit at a time. *Datapin* can be 0 to 15 or a *Portname.number*. *\bits* optionally specify the number of bits to be shifted out and if omitted, 8 bits are assumed. *Mode* specifies which bit will be sent out first. If *Mode* is 0, the LSB is sent out first followed by other data bits. If *Mode* is 1, the MSB is sent out first followed by other data bits.

In the following example, the contents of variable B1 are sent out as 8 bits, LSB first, from bit 0 of PORTB. Bit 1 of PORTB is used as the clock pin.

> SHIFTOUT PORTB.0, PORTB.1, 0, B1

SWAP

> SWAP *Var, Var*

This command is used to swap the contents of two variables. It can be used with bit, byte, and word variables.

In the following examples, values of variables B1 and B2 are exchanged:

> SWAP B1, B2

WHILE..WEND

> WHILE *condition*
> > Statement...
> WEND

This is another command used to create loops in programs. The statements between the WHILE and WEND are repeated while the *Condition* is true.

In the following example, the statements between WHILE and WEND are repeated 10 times:

```
k = 0
WHILE k < 10
    Sum = Sum + 1
    B0 = B0 + 2
    k = k + 1
WEND
```

4.3   Liquid crystal display (LCD) interface and commands

In many microcontroller-based applications, it is required to display a message or the value of a variable. For example, in a temperature-control application, it may be required to display the value of the temperature dynamically. Basically, three types of displays can be used in practise. These are video displays, 7-segment LED displays, and LCD displays. Standard video displays require complex interfaces and their cost is relatively high and their operation is not covered in this book. 7-segment LED displays are made up of LEDs. Although the 7-segment LEDs are bright, their disadvantage is the high power consumption which makes them unsuitable in many battery-operated portable applications. We will see the operation of these devices in Chapter 5.

LCDs are alphanumeric displays which are frequently used in microcontroller-based applications. Some of the advantages of LCDs are their low cost and low power consumption. LCDs are ideal in low-power, battery-operated portable applications. These displays come in different shapes and sizes. Some LCDs have 40 or more characters with several rows. Some more advanced LCDs can be programmed to display graphics images. Some modules, such as the ones used in games, offer colour displays while some others may incorporate back lighting so that they can be viewed in dimly lit conditions. In this section, we shall be looking at how we can interface the standard LCDs to a PIC microcontroller and what commands are available to use the LCDs.

There are basically two types of LCDs as far as the interface technique is concerned: parallel LCDs and serial LCDs. Parallel LCDs are connected to the microcontroller I/O ports using 4 or 8 data wires and data is transferred from the microcontroller to the LCD in parallel form. Serial LCDs are connected to the microcontroller using only one data line and data is transferred to the LCD using the standard RS232 asynchronous data communication protocols. Serial LCDs are easier to use but they usually cost more than the parallel ones. Serial LCDs also have the advantage that only one wire is required to interface them to a microcontroller, thus saving the I/O pins. In this section, we shall be looking at the interface and programming of both types of LCDs.

4.3.1   Parallel LCDs

Figure 4.8 shows a typical parallel LCD. The programming of a parallel LCD is usually a complex task and requires a good understanding of the internal operation of the LCDs, including the timing requirements. Fortunately, the PicBasic Pro language provides special commands for displaying data on HD44780 or compatible LCDs. All the user has to do is connect the LCD to the appropriate I/O ports and then use these special commands to simply send data to the LCD. The standard

Figure 4.8   A typical parallel LCD

PicBasic language does not provide any special commands for programming the parallel LCDs and the programming of LCDs using the PicBasic language is described in the projects section (Chapter 5) of this book.

HD44780 LCD module

HD44780 is one of the most popular LCD modules used in the industry and also by hobbyists. This module is monochrome and comes in different shapes and sizes. Modules with line lengths of 8, 16, 20, 24, 32, and 40 characters can be selected. Depending on the model chosen, 1, 2, or 4 display rows can be selected. The display has a 14-pin connector for interfacing to a microcontroller. Table 4.11 shows the pin configuration of the LCD. A description of the pin functions is given below.

- V_{SS} is the 0 V or ground. V_{DD} pin should be connected to the positive supply. Although the manufacturers specify a 5 V supply, the module can be operated with as low as 3 V or as high as 6 V.
- Pin 3 is named as V_{EE} and this is the contrast control pin. This pin is used to adjust the contrast of the LCD and it should be connected to a variable voltage supply. A potentiometer is usually connected between the power supply lines with its wiper arm connected to this pin so that the contrast can be adjusted. This pin can be connected to ground if contrast adjustment is not needed.
- Pin 4 is the Register Select (RS) and when this pin is LOW, data transferred to the display is treated as commands. When RS is HIGH, character data can be transferred to and from the module.
- Pin 5 is the read/write (R/W) pin. This pin is pulled LOW in order to write commands or character data to the LCD module. When this pin is HIGH, character data or status information cannot be read from the module. This pin is usually connected to ground, i.e. the LCD is put into write mode.
- Pin 6 is the Enable (E) pin which is used to initiate the transfer of commands or data between the LCD module and the microcontroller. When writing to the display, data is transferred only on the

Table 4.11   Pin configuration of HD44780 LCD

Pin No	Name	Function
1	V_{SS}	Ground
2	V_{DD}	Positive supply
3	V_{EE}	Contrast
4	RS	Register select
5	R/W	Read/write
6	E	Enable
7	D0	Data bit 0
8	D1	Data bit 1
9	D2	Data bit 2
10	D3	Data bit 3
11	D4	Data bit 4
12	D5	Data bit 5
13	D6	Data bit 6
14	D7	Data bit 7

HIGH to LOW transition of this pin. When reading from the display, data becomes available after the LOW to HIGH transition of the enable pin and this data remains valid as long as the enable pin is HIGH.

• Pins 7 to 14 are the eight data bus lines (D0 to D7). Data can be transferred between the micro-controller and the LCD module using either an 8-bit interface, or a 4-bit interface. In the latter case, only the upper four data lines (D4 to D7) are used and the data is transferred as two 4-bit nibbles. This mode has the advantage that fewer I/O lines are required to communicate with the LCD.

Connecting the LCD to the microcontroller

PicBasic Pro compiler by default assumes that the LCD is connected to specific pins of the micro-controller unless told otherwise. It assumes the following connections:

LCD	Microcontroller
D4	RA0
D5	RA1
D6	RA2
D7	RA3
E	RB3
RS	RA4

Figure 4.9 shows the circuit diagram with the default connections between the LCD and the microcontroller. In addition to the above connections, the R/W pin of the LCD is not used and is connected to the ground. The contrast adjustment is done by connecting a potentiometer to V_{EE}. Notice that port pin RA4 is connected to +5 V supply with a resistor. This is because this pin is open-drain output and should be pulled HIGH with a resistor.

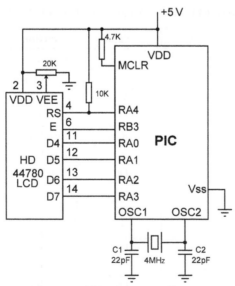

Figure 4.9   Default LCD connections to a PIC microcontroller

When the above connections are made between the microcontroller and the LCD, we can simply use the LCDOUT command to send data to the LCD module. Note that the connections between the microcontroller and the LCD can be changed using a set of DEFINE commands to assign the LCD pins to the PIC microcontroller.

In the following example, PORTB pins 0 to 4 are used for LCD data (i.e. RB0 connected to D4, RB5 connected to D5, etc.), bit 4 of PORTB is connected to the RS pin of the LCD, bit 5 of PORTB is connected to the E pin of the LCD, the LCD is set for 4-bits of operation, and the LCD is assumed to have two rows.

```
DEFINE LCD_DREG      PORTB    ' Set LCD data port to PORTB
DEFINE LCD_DBIT      0        ' Set data starting bit to 0
DEFINE LCD_RSREG     PORTB    ' Set RS register port to PORTB
DEFINE LCD_RSBIT     4        ' Set RS register bit to 4
DEFINE LCD_EREG      PORTB    ' Set E register port
DEFINE LCD_EBIT      5        ' Set E register bit to 5
DEFINE LCD_BITS      4        ' Set 4 bit operation
DEFINE LCD_LINES     2        ' Set number of LCD rows
```

The format of the LCDOUT command is

   LCDOUT *Item, Item,......*

where *Item* can be a command or data. A command is used to clear the display, home the cursor, move the cursor to left or right, etc. It is important that a program should wait for at least half a second before sending the first command to the LCD. This is because it can take quite a while before the LCD initializes itself.

Table 4.12 gives a list of the available commands. All commands must be preceded by the hexadecimal number $FE. For example, to clear the display we have to issue the command

   LCDOUT $FE, 1

Similarly, to move the cursor left by one position we have to issue the command

   LCDOUT $FE, $10

Also, to move the cursor to the 5th position in the first row, we have to use the command

   LCDOUT $FE, $80 + 5

Data is sent to the LCD using the LCDOUT command. The character set of the LCD is given in Table 4.13. A string can be sent to the LD by enclosing it in double-quotes. For example, the following command displays the string HELLO at the current cursor position:

   LCDOUT "HELLO"

Table 4.12   LCD commands

Command	Operation
$FE, 1	Clear display
$FE, 2	Home cursor
$FE, $0C	Cursor off
$FE, $0E	Underline cursor on
$FE, $0F	Blinking cursor on
$FE, $10	Move cursor left by one position
$FE, $14	Move cursor right by one position
$FE, $80	Move cursor to the beginning of first row
$FE, $C0	Move cursor to the beginning of second row
$FE, $94	Move cursor to the beginning of third row
$FE, $D4	Move cursor to the beginning of fourth row

If a hash sign (#) precedes a variable (or if the characters DEC precede a variable), the ASCII representation for each digit is sent to the LCD. For example, if the variable B1 = 208, then the command

LCDOUT #B1

or

LCDOUT DEC B1

displays the characters "2", "0", and "8" on the LCD.

If character BIN precedes a variable, the ASCII representation of its binary value is sent to the LCD. For example, if the variable B1 = 9, then the command

LCDOUT BIN B1

displays the characters "1001" on the LCD.

A numeric value preceded by HEX will send the ASCII representation of its hexadecimal value to the LCD. For example, if B0 = 255, then the command

LCDOUT HEX B0

will display "FF" on the LCD.

It is also possible to send repeated characters to the LCD. In the following example, the characters "AAAAA" are sent to the LCD:

LCDOUT REP "A"\5

Table 4.13 LCD character table

Lower 4 Bits \ Upper 4 Bits	0000	0001	0010	0011	0100	0101	0110	0111	1000	1001	1010	1011	1100	1101	1110	1111
xxxx0000	CG RAM (1)			0	@	P	`	p				―	９	ミ	α	p
xxxx0001	(2)		!	1	A	Q	a	q			。	７	チ	ム	ä	q
xxxx0010	(3)		"	2	B	R	b	r			「	イ	ツ	メ	β	θ
xxxx0011	(4)		#	3	C	S	c	s			」	ウ	テ	モ	ε	∞
xxxx0100	(5)		$	4	D	T	d	t			、	エ	ト	ヤ	μ	Ω
xxxx0101	(6)		%	5	E	U	e	u			・	オ	ナ	ユ	σ	Ü
xxxx0110	(7)		&	6	F	V	f	v			ヲ	カ	ニ	ヨ	ρ	Σ
xxxx0111	(8)		'	7	G	W	g	w			ア	キ	ヌ	ラ	g	π
xxxx1000	(1)		(	8	H	X	h	x			ィ	ク	ネ	リ	√	x̄
xxxx1001	(2)		)	9	I	Y	i	y			ゥ	ケ	ノ	ル	‐‐	y
xxxx1010	(3)		*	:	J	Z	j	z			エ	コ	ハ	レ	j	千
xxxx1011	(4)		+	;	K	[	k	{			オ	サ	ヒ	ロ	×	万
xxxx1100	(5)		,	<	L	¥	l	|			ヤ	シ	フ	ワ	¢	円
xxxx1101	(6)		―	=	M	]	m	}			ュ	ス	ヘ	ン	Ł	÷
xxxx1110	(7)		.	>	N	^	n	→			ョ	セ	ホ	゛	ñ	
xxxx1111	(8)		/	?	O	_	o	←			ッ	ソ	マ	°	Ö	■

Example 4.1

A 2-row parallel LCD is connected to a PIC microcontroller as shown in Figure 4.9. Write a PicBasic Pro program to display the string "PIC ROW 1" and "PIC ROW 2" in row 1 and row 2 of the LCD, respectively.

Solution 4.1

The required program is

```
        PAUSE 1000            ' Wait 1 second for initialization
        LCDOUT $FE ,1         ' Clear the LCD

        LCDOUT "PIC ROW 1"    ' Display message in row 2
        LCDOUT $FE, $C0       ' Move cursor to row 2
        LCDOUT "PIC ROW 2"    ' Display message in row 2
```

4.3.2 Serial LCDs

A serial LCD is connected to a microcontroller using only one data line. Both PicBasic and PicBasic Pro languages can be used to send data to serial LCDs using the SEROUT command.

A popular serial LCD is the ILM-216 (see Figure 4.10). This is a 16-pin, 2-row by 16-character LCD manufactured by Scott Edwards Electronics Inc. The device can operate with a baud rate of 2400 or 9600. In addition to the normal display functions, inputs for four push-button switches and also an output to drive a buzzer are included on the LCD module. The module incorporates an EEPROM memory and a backlight which are programmable.

Figure 4.10   ILM-216 serial LCD

Table 4.14 shows the pin configuration of this LCD. Pins 1 and 2 are the ground and the +5 V supply connections, respectively. Pin 3 is the serial input pin. Either RS232 voltage levels or standard TTL level signals can be connected to this pin. Similarly, pin 4 is the serial output pin and TTL logic levels (inverted) can be connected to this pin. Pin 5 is the buzzer out pin where a small buzzer (up to 25 mA) can be connected to this pin and the buzzer can be controlled with the software. Pins 6 to 8 are the option pins. Pin 7 is used to configure the device. Pin 8 is used to select a baud rate and when this pin is connected to pin 6, the device operates at 9600 baud. Leaving pin 8 unconnected configures the device to operate at 2400 baud. Pins 9 to 16 are four push-button switch inputs. The state of these pins can be read from the software.

Table 4.14 Pin configuration of ILM-216

Pin No	Function
1	Ground
2	+5 V
3	Serial in
4	Serial out
5	Bell
6	Ground
7	Config/test
8	9600 baud
9	Switch 1
10	Switch 1 ground
11	Switch 2
12	Switch 2 ground
13	Switch 3
14	Switch 3 ground
15	Switch 4
16	Switch 4 ground

The ILM-216 can be connected to a microcontroller using the following minimum pins:

Pin 1   ground

Pin 2   +5 V supply

Pin 3   to microcontroller serial output

Pin 4   to microcontroller serial input (if it is required to read the state of push-button switches on the LCD module)

The default factory configuration of the ILM-216 is 2400 baud, 8 data bits, no parity, and 1 stop bit. Table 4.15 gives a list of the control codes of ILM-216. These codes are summarized below:

Null: These characters are ignored by the LCD

Table 4.15   ILM-216 LCD control codes

Function	ASCII Code
Null	0
Cursor home	1
Hide cursor	4
Show underline cursor	5
Show blinking cursor	6
Bell	7
Backspace	8
Horizontal tab	9
Smart line feed	10
Vertical tab	11
Clear screen	12
Carriage return	13
Backlight on	14
Backlight off	15
Cursor position	16
Format right-aligned text	18
Escape codes	27

Cursor home: Moves the cursor to the first character position of the first row

Hide cursor: Hides the cursor so that it is not visible

Show underlined cursor: Shows a non-blinking underlined cursor at the current position

Show blinking cursor: Shows a blinking cursor at the current position

Bell: sends pulses to a buzzer connected to pin 5 of the LCD

Backspace: Moves the cursor back by one space and erases the character in that position

Smart line feed: Moves the cursor down by one line

Vertical tab: Moves the cursor up by one line

Clear screen: Clears the LCD screen

Carriage return: Moves the cursor to the first position on the next row

Backlight on: Turns on the LED backlight

Backlight off: Turns off the LED backlight

Position cursor: Accepts a number from 0 to 31 and moves the cursor to that position where 0 is the first character of the first row and 31 is the last character of the second row. Number 64 should be added to the required cursor position in order to get the actual displayed cursor position. For example, position 80 corresponds to the first character position in the second row (64 + 16 = 80).

Right align text: Accepts a number from 2 to 9 representing the width of an area on the screen in which right-aligned text is to be displayed.

Escape sequences: Escape codes enable the user to define a custom character, to transfer data from the EEPROM, and to read the state of the four push-button switch positions on the LCD module.

Example 4.2

An ILM-216 model serial LCD is connected to bit 0 of PORTB of a PIC microcontroller as shown in Figure 4.11. Write a PicBasic Pro program to clear the LCD screen and then to display the string "PIC LCD" in row 1 of the LCD. Wait 1 s for the initialization of the LCD.

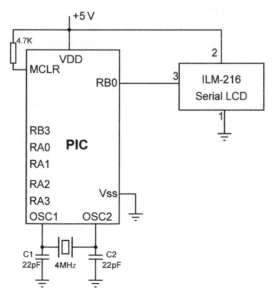

Figure 4.11   Connecting ILM-216 model LCD to a PIC microcontroller

Solution 4.2

The required program is given below. The PicBasic command SEROUT is used to send data to the serial LCD.

```
PAUSE 1000              ' Wait 1 s for initialization
SEROUT PORTB.0, N2400 (12, "PIC LCD")
```

4.4   Interrupts

Interrupts are very useful in many microcontroller applications. An interrupt, as the name suggests, interrupts the normal execution of a program and jumps to a designated address in the program memory called the *Interrupt Service Routine* (ISR) where a short program is executed. At the end of this program, control is returned to the main program and execution continues from the point it was interrupted.

Interrupts are asynchronous events and it is not known when they may occur. There are basically two types of interrupts: external interrupts and internal interrupts. External interrupts may occur when an external event occurs. For example, when an external signal changes its state. Internal interrupts are usually in the form of timer interrupts and an interrupt may be generated when the timer overflows.

When an interrupt occurs, the PIC microcontroller saves the address of the next instruction on stack and jumps to the ISR which is at address 4 of the program memory. When interrupts are expected from multiple sources, the program should check at the beginning of the ISR to determine the actual source of the interrupt.

PicBasic Pro allows the use of interrupts in programs. The command

ON INTERRUPT GOTO *Label*

declares *Label* as the starting point of the ISR. Further interrupts should be disabled by the DISABLE command just before entering the ISR. Also, further interrupts should be enabled by the ENABLE command after the end of the ISR. The last statement in the ISR should be the RESUME statement which terminates the ISR and returns to the main program.

The structure of the main program and the ISR are as follows:

Main program

```
ON INTERRUPT GOTO Mylabel
........................................................
........................................................
........................................................
```

Interrupt Service Routine

DISABLE
Mylabel:

....

....

....

RESUME
ENABLE

The use of external and timer interrupts will be discussed further with examples in the projects section of this book.

4.5   Recommended PicBasic Pro program structure

A PicBasic Pro program can be written in many different formats. The author recommends that you use a template similar to the one given in Figure 4.12 when developing PicBasic Pro programs. As you can see in this figure, the header includes a brief description of the program including the author name, the date, and filename of the program. Comments are used in every line of the program to clarify the actions of the program.

```
'*******************************************************************

'

'                        LED FLASHING PROGRAM
'                        =========================

'
' This program flashes and LED connected to port RB0 of PORTB.
' The Led is flashed with 1 second intervals.

' Author:        Dogan Ibrahim
' Date:          September, 2005
' File:          LED.PBP
'

' Modifications
' =============


'
'*******************************************************************
```

Figure 4.12   (Continued)

```
'
' DEFINITIONS
'
LED VAR PORTB.0                          ' Define RB0 as LED


'
' START OF MAIN PROGRAM
'
        TRISB = 0                        ' Set PORT B pins as outputs

AGAIN:
        LED = 1                          ' Turn ON LED
        PAUSE 1000                       ' Wait 1 second

        LED = 0                          ' Turn OFF LED
        PAUSE 1000                       ' Wait 1 second

        GOTO AGAIN                       ' Repeat

        END                              ' End of program
```

Figure 4.12   Recommended PicBasic Pro program template

4.6   Using stepping motors

Stepping motors are widely used in many microcontroller-based projects where motion is required. This section describes the basic operation of these motors and also shows how they can be used in microcontroller-based projects with PicBasic and PicBasic Pro languages.

Stepping motors are electro-mechanical devices that convert electrical pulses into discrete mechanical movements. A conventional motor has a free running shaft and rotates continuously as long as power is applied to the motor. The shaft of a stepping motor rotates in discrete steps when electrical pulses are applied to it in the correct sequence. The speed of the rotation is related to the time between the input pulses and the length of rotation is directly related to the number of pulses applied. Basically, the motor rotates by an angle defined as the "stepping angle" each time a pulse is applied to the motor. For example, if the stepping angle of a stepping motor is specified as 10°, then each time a pulse is applied the motor will rotate by an angle of 10° and 36 pulses will be required to make a complete 360° rotation.

Stepping motors have the following advantages over the conventional motors:

- Motor shaft position can be controlled very accurately using pulses and in open-loop mode.
- Stepping motors can be operated at very low speeds.

- Stepping motors are very reliable since there are no brushes and, as a result, these motors have very long operational lives.
- Stepping motors have full torque at standstill.
- The speed of stepping motors can be controlled easily and accurately.

There are basically two types of stepping motors: unipolar and bipolar. Unipolar motors are easy to control where two windings with common points are used, and a simple 1-of-*n* counter circuit can be used to generate the required stepping sequence. A driver transistor can be used for each winding. One of the most commonly used drive methods is 1 phase full step, also known as the "wave drive", where the motor windings are energised one at a time as shown in Table 4.16. The motor can be driven by using a MOSFET power transistor for each coil winding, as shown in Figure 4.13. Unipolar motors can also be driven by using integrated circuits, such as the UCN5804B. This chip operates with voltages between 6 and 30 V. It contains a CMOS logic section for the sequencing logic and a high-voltage output section to directly drive a unipolar stepping motor. As shown in Figure 4.14, the motor is connected directly to the chip and the chip

Table 4.16   One-phase full-step drive

Step	A	B	C	D
1	1	0	0	0
2	0	1	0	0
3	0	0	1	0
4	0	0	0	1

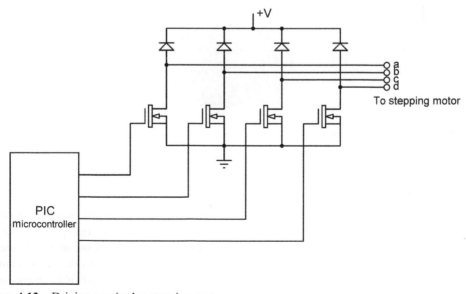

Figure 4.13   Driving a unipolar stepping motor

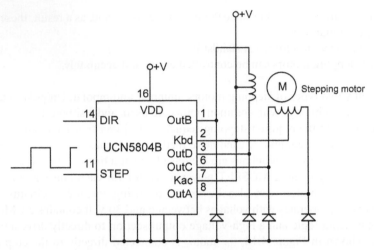

Figure 4.14   Controlling a unipolar motor using a UCN5804B

generates the correct sequence of signals to drive the motor. The DIR input controls the direction of rotation. The motor is rotated by one step each time a pulse is applied to the STEP input.

Bipolar motors generally produce higher torques, but more complex circuits are required to control these motors. The control of bipolar stepping motors is beyond the scope of this book.

Figure 4.15 shows a typical small stepping motor.

Figure 4.15   A typical stepping motor

4.7   Using servomotors

Servomotors are generally used in radio control toys, such as airplanes, boats, or robots. A servomotor consists of a DC motor with a series of gears attached to it. An internal potentiometer is used with feedback to control the movement of the motor. Normally, the output shaft is limited to 180° of rotation, but it is possible to modify a servomotor so that continuous rotation is obtained. In the projects section of this book, we shall be looking at the control of modified servomotors.

A servomotor is controlled with pulse-width-modulated (PWM) signal. In a modified servo-motor, a pulse with a width of 2 ms rotates the motor clockwise at full speed. Similarly, a pulse with a width of 1 ms rotates the motor anti-clockwise at full speed. Sending a pulse with a width of 1.5 ms stops the motor.

A servomotor requires only three wires to operate: +V, ground, and the signal wire where the pulse is applied.

Figure 4.16 shows a typical small servomotor.

Figure 4.16   A typical servomotor

4.8   Exercises

1. What are the ranges of PicBasic variables bit, byte and word?
2. Explain how you can declare a 20-element byte array called *scores* in PicBasic.
3. Explain how you can use comments in PicBasic and PicBasic Pro programs. Why should we use comments in our programs?
4. Why would you use *Symbols* in a PicBasic program?
5. Explain the use of the command BRANCH by giving an example.
6. Give different ways in which you can make loops in PicBasic Pro programs.
7. Explain how you can connect an LCD to a PIC microcontroller using the default settings.
8. Write a program to count from 0 to 100 repeatedly with 1 s intervals and show your results on a parallel LCD.
9. Explain the advantages and the disadvantages of parallel and serial LCDs.
10. Write a PicBasic Pro program to display the text "RESULTS" in row 1, column 5 of a parallel LCD.
11. Write a PicBasic Pro program to count from 0 to 100 in steps of 2 and show the output on the second row of a parallel LCD.
12. Repeat question 11 using a serial LCD and PicBasic language.
13. Explain how you can use the DEFINE statements to change the interface between a PIC microcontroller and an LCD.

14. Give an example for the use of the SELECT..CASE command. Show how you can program using the IF..THEN..ELSE command instead. Explain which one you would prefer.
15. Explain the differences between the REPEAT..UNTIL and WHILE..WEND commands. Give examples for each command.
16. Explain how a unipolar stepping motor can be controlled.
17. Explain how a modified servomotor can be controlled to rotate: (a) full speed clockwise, (b) full speed anti-clockwise.

5
PicBasic and PicBasic Pro projects

In previous chapters we have seen the characteristics of the PIC microcontrollers and how to program these microcontrollers using the PicBasic and PicBasic Pro languages. In this chapter we shall be looking at various PIC microcontroller-based projects. All the projects described here have been constructed and tested using both the PicBasic and PicBasic Pro languages.

Each project has been described with the following sub-headings:

Project title:	Title of the project
Project description:	A brief description of the project.
Hardware:	Hardware used in the project. This is mainly the circuit diagram of the microcontroller and associated interface electronics used for the project.
Flow diagram:	A flow diagram is given to describe the operation of the project.
Software:	Listings of the microcontroller programs for both PicBasic and PicBasic Pro languages.

Projects in this chapter have been organised in increasing complexity. It is recommended that the reader study the simple projects first before going to the more complex ones.

Project 1

Project title: Simple flashing LED

Project description: An LED is connected to one of the port pins of a PIC microcontroller. The LED is flashed continuously with 1-s interval.

Hardware: This project is so simple that any type of PIC microcontroller can be used. As shown in Figure 5.1, a PIC16F84 type microcontroller is chosen for this project. Bit 0 of PORTB (RB0) is connected to a small LED through a current-limiting resistor. The voltage drop across an LED is approximately 2 V. Assuming an LED current of 10 mA, the value of the resistor can be calculated as

$$R = \frac{V}{I} = \frac{5 - 2}{10\,\text{mA}} = 0.3\,\text{K}$$

the nearest value is 330 Ω.

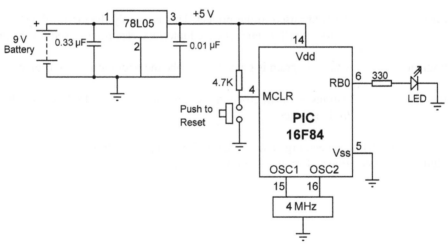

Figure 5.1   Circuit diagram of Project 1

The project has been constructed on a breadboard as shown in Figure 5.2.

Flow diagram: The software consists of an indefinite loop where the LED is turned on and off inside this loop. The flow diagram of the software is shown in Figure 5.3.

Figure 5.2   Construction of Project 1

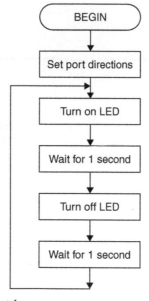

Figure 5.3   Flow diagram of Project 1

Software: **PicBasic**
                       The software for PicBasic language is shown in Figure 5.4. At the begin-
                       ning of the program LED is defined as a symbol and is assigned to zero
                       (bit 0 of PORTB). Also, the port direction register TRISB and PORTB

addresses are defined. The main program is an indefinite loop and starts with label AGAIN. Inside the main program the LED is turned on using the HIGH LED instruction. Then after a delay of 1 s (PAUSE 1000) the LED is turned off and this process is repeated forever.

```
'**********************************************************
'
'               LED FLASHING PROGRAM
'               =======================
'
' This program flashes an LED connected to port RB0 of PORTB. The
' Led is flashed with 1 second intervals.
'
' Author:      Dogan Ibrahim
' Date:        October, 2005
' Compiler:    PicBasic
' File:        LED1.BAS
'
' Modifications
' ===========
'
'**********************************************************
'
' SYMBOLS
'
Symbol LED = 0                          ' Define RB0 as LED
Symbol TRISB = $86                      ' TRISB address
Symbol PORTB = $06                      ' PORTB address
'
' START OF MAIN PROGRAM
'
        POKE TRISB, 0                   ' Set PORTB pins as outputs

AGAIN:
        HIGH LED                        ' Turn ON LED
        PAUSE 1000                      ' Wait 1 second

        LOW LED                         ' Turn OFF LED
        PAUSE 1000                      ' Wait 1 second

        GOTO AGAIN                      ' Repeat

        END                             ' End of program
```

Figure 5.4   PicBasic program of Project 1

PicBasic Pro

The software for PicBasic Pro language is shown in Figure 5.5. At the beginning of the program LED is defined as bit 0 of PORTB (PORTB.0). Port direction register TRISB is then cleared so that all PORTB pins are outputs. Main program starts with label AGAIN where the port pin is turned on and off with 1 s intervals.

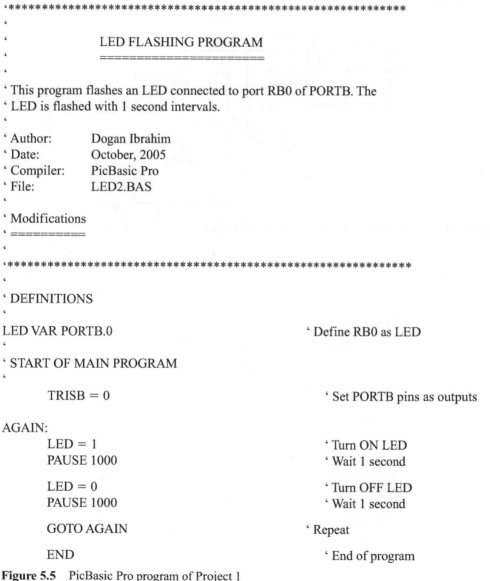

```
'************************************************************
'
'              LED FLASHING PROGRAM
'              =========================
'
' This program flashes an LED connected to port RB0 of PORTB. The
' LED is flashed with 1 second intervals.
'
' Author:      Dogan Ibrahim
' Date:        October, 2005
' Compiler:    PicBasic Pro
' File:        LED2.BAS
'
' Modifications
' ===========
'
'************************************************************
'
' DEFINITIONS
'
LED VAR PORTB.0                              ' Define RB0 as LED
'
' START OF MAIN PROGRAM
'
        TRISB = 0                                ' Set PORTB pins as outputs

AGAIN:
        LED = 1                                  ' Turn ON LED
        PAUSE 1000                               ' Wait 1 second

        LED = 0                                  ' Turn OFF LED
        PAUSE 1000                               ' Wait 1 second

        GOTO AGAIN                           ' Repeat

        END                                      ' End of program
```

Figure 5.5   PicBasic Pro program of Project 1

Using a different microcontroller

In this project a PIC16F84-type microcontroller has been used. Recently, PIC16F627 has become one of the popular low-cost PIC microcontrollers. This is an 18-pin microcontroller, pin compatible with the PIC16F84, having 16 I/O ports and built-in 4-MHz-clock oscillator. In this section we shall be using the PIC16F627 to flash the LED.

Figure 5.6 shows the circuit diagram of the PIC16F627-based project. The LED is connected to bit 0 of PORTB as in Figure 5.1 and the internal oscillator of the microcontroller is used.

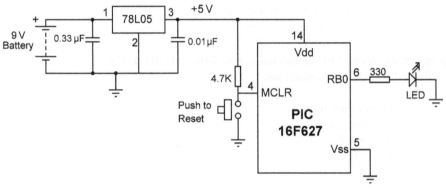

Figure 5.6   Circuit diagram of the PIC16F627-based project

Figure 5.7 shows the construction of the project on a breadboard. Notice that there are no timing components in this circuit.

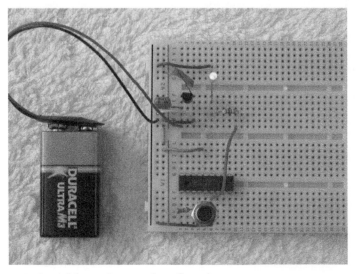

Figure 5.7   Construction of the project on a breadboard

PicBasic and PicBasic Pro programs of the project are same as in Figures 5.4 and 5.5, respectively. The internal 4-MHz-clock oscillator should be selected during programming of the microcontroller as shown in Figure 5.8.

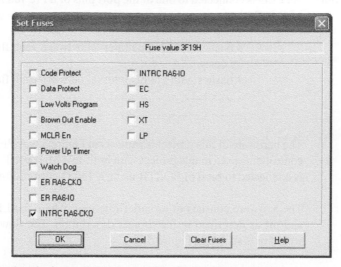

Figure 5.8   Selecting the internal 4 MHz oscillator during programming

Project 2

Project title: Complex flashing LED

Project description: An LED is connected to one of the port pins of a PIC microcontroller. The LED is flashed continuously as in the following sequence:

> 3 flashes with 250ms interval between each flash.
> 2 s delay.
> 3 flashes with 250ms interval between each flash.
>
> ....
>
> ....

Hardware: The hardware of this project is same as in Figure 5.6. A PIC16F627 microcontroller is used in this project with built-in 4 MHz oscillator and an LED is connected to bit 0 of PORTB using a 330 Ω current-limiting resistor.

Flow diagram: The software consists of an indefinite loop where the LED is turned on and off as described in the project description. The flow diagram of the software is shown in Figure 5.9.

Software: **PicBasic**
                         The software for PicBasic language is shown in Figure 5.10. At the beginning of the program LED is defined as a symbol and is assigned to zero (bit 0 of PORTB). Also, the port-direction register TRISB and PORTB addresses are defined. The main program is an indefinite loop and starts with label AGAIN. Inside the main program a OR loop is formed and the LED is flashed three times with 250 ms intervals. After a 2 s delay the process is repeated. Variable *Cnt* is used as the loop-count variable.

                         PicBasic Pro
                         The software for PicBasic Pro language is shown in Figure 5.11. At the beginning of the program port-direction register TRISB is cleared so that all PORTB pins are outputs. Main program starts with label AGAIN. Inside the main program a FOR loop is formed and the LED is flashed three times with 250 ms intervals. After a 2 s delay the process is repeated. Variable *Cnt* is used as the loop-count variable.

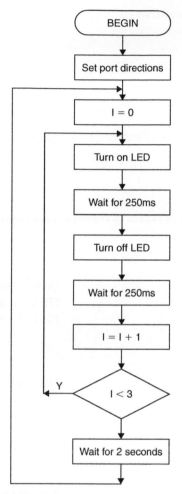

Figure 5.9 Flow diagram of Project 2

```
'****************************************************************
'
'                  LED FLASHING PROGRAM
'                  =====================
'
' This program flashes an LED connected to port RB0 of PORTB. The
' LED is flashed continuously as follows:
'
'          Flash 3 times with 250ms intervals
'          Wait 2 seconds
```

Figure 5.10   (Continued)

```
'        Flash 3 times with 250ms intervals
'        ...............................................
'        ...............................................
'
'
' Author:      Dogan Ibrahim
' Date:        October, 2005
' Compiler:    PicBasic
' File:        LED3.BAS
'
' Modifications
' ===========
'
'*******************************************************************
'
' SYMBOLS
'
Symbol LED = 0                          ' Define RB0 as LED
Symbol TRISB = $86                      ' TRISB address
Symbol PORTB = $06                      ' PORTB address
'
' VARIABLES
'
Symbol Cnt = B0                         ' Declare Cnt as a byte
'
' START OF MAIN PROGRAM
'
        POKE TRISB, 0                   ' Set PORTB pins as outputs
AGAIN:
        FOR Cnt = 1 TO 3
            HIGH LED                    ' Turn ON LED
            PAUSE 250                   ' Wait 250ms
            LOW LED                     ' Turn OFF LED
            PAUSE 250                   ' Wait 250ms
        NEXT Cnt

        PAUSE 2000                      ' Wait 2 seconds

        GOTO AGAIN                      ' Repeat

        END                             ' End of program
```

Figure 5.10   PicBasic program of Project 2

```
'****************************************************************
'
'                    LED FLASHING PROGRAM
'                    ========================
'
' This program flashes an LED connected to port RB0 of PORTB. The
' LED is flashed continuously as follows:
'
'          Flash 3 times with 250ms intervals
'          Wait 2 seconds
'          Flash 3 times with 250ms intervals
'          ................................................
'          ................................................
'
'
' Author:      Dogan Ibrahim
' Date:        October, 2005
' Compiler:    PicBasic Pro
' File:        LED4.BAS
'
' Modifications
' ===========
'
'****************************************************************
'
'
' DEFINITIONS
'
Cnt   VAR   BYTE                        ' Declare Cnt as a byte
'
' START OF MAIN PROGRAM
'
    TRISB = 0                           ' Set PORTB pins as outputs
AGAIN:
    FOR Cnt = 1 TO 3
        PORTB.0 = 1                     ' Turn ON LED
        PAUSE 250                       ' Wait 250ms
        PORTB.0 = 0                     ' Turn OFF LED
        PAUSE 250                       ' Wait 250ms
    NEXT Cnt
    PAUSE 2000                          ' Wait 2 seconds
    GOTO AGAIN                          ' Repeat
    END                                 ' End of program
```

Figure 5.11   PicBasic Pro program of Project 2

Project 3

Project title: Flashing LED warning lights

Project description: In this project, two LEDs are connected to bit 0 of PORTB of a PIC
                           microcontroller. The LEDs turn on and off alternately with 1 s delay.

Hardware: The hardware of this project is similar to the circuit given in Figure 5.6.
                           But here, two LEDs are connected to the same output pin of the microcon-
                           troller. When the pin output is logic 1, the microcontroller is sourcing cur-
                           rent and the lower LED is turned on and the upper LED is off. Similarly,
                           when the pin output is logic 0, the microcontroller is sinking current and
                           the upper LED is turned on and the lower LED is off. 330 Ω current-limiting
                           resistors are used for each LED. The circuit diagram of the project is
                           shown in Figure 5.12.The construction of the project on a breadboard is
                           shown in Figure 5.13.

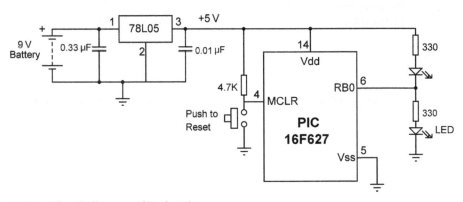

Figure 5.12   Circuit diagram of Project 3

Flow diagram: The flow diagram of the project is as in Figure 5.3, i.e. the output pin of
                           the microcontroller is turned on and off with 1 s intervals.

Software: **PicBasic**
                           The software for PicBasic language is exactly same as given in Figure 5.4.

                           PicBasic Pro
                           The software for PicBasic Pro language is exactly same as given in Figure 5.5.

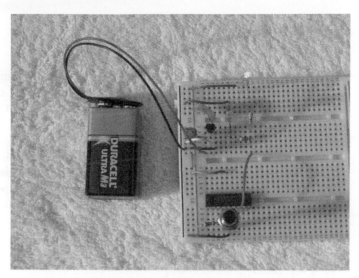

Figure 5.13   Construction of the project on a breadboard

Project 4

Project title: Turning on odd numbered LEDs

Project description: In this project, 8 LEDs are connected to PORTB of a PIC microcontroller. When the project is started (or when reset), only the odd numbered LEDs turn on (i.e. the LEDs connected to bit 1, bit 3, bit 5, and bit 7 of PORTB).

Hardware: The circuit diagram of the project is shown in Figure 5.14. A PIC16F627 model PIC microcontroller is used and the microcontroller is operated from its 4 MHz internal clock. The LEDs are connected to 8 pins of PORTB using 330 Ω current-limiting resistors. An external reset button is connected to MCLR input of the microcontroller.

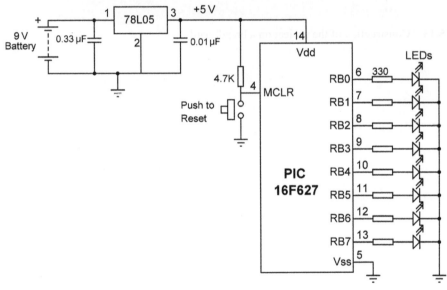

Figure 5.14   Circuit diagram of Project 4

The construction of the project on a breadboard is shown in Figure 5.15.

Flow diagram: The flow diagram of the project is shown in Figure 5.16. At the beginning of the program the I/O direction is specified. And then the hexadecimal number $AA is sent to PORTB to turn on the odd-numbered LEDs. Note that

$$\$AA = 10101010$$

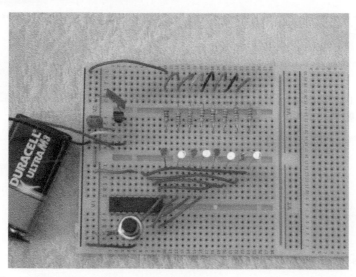

Figure 5.15   Construction of the project on a breadboard

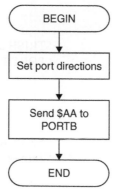

Figure 5.16   Flow diagram of Project 4

i.e. the odd numbered bit positions are logic 1, and even-numbered bit positions are logic 0.

Software:
    PicBasic
The software for PicBasic language is given in Figure 5.17. At the beginning of the program PORTB and TRISB addresses are defined. TRISB is then cleared to 0 to make all PORTB pins as outputs. Then the hexadecimal number $AA is sent to PORTB using the POKE statement to turn on the odd-numbered LEDs.

```
'****************************************************************
'
'                 TURN ON ODD NUMBERED LEDS
'                 ===============================
'
' This program turns on odd numbered LEDs (bit 1, bit 3, bit 5, bit 7) connected
' to PORTB of a PIC16F627 microcontroller.
'
'
'
' Author:        Dogan Ibrahim
' Date:          October, 2005
' Compiler:      PicBasic
' File:          LED5.BAS
'
' Modifications
' =============
'
'****************************************************************
'
' SYMBOLS
'
Symbol TRISB = $86                     ' TRISB address
Symbol PORTB = $06                     ' PORTB address


'
' START OF MAIN PROGRAM
'
      POKE TRISB, 0                    ' Set PORTB pins as outputs
      POKE PORTB, $AA                  ' Turn on odd numbered LEDs

      END                              ' End of program
```

Figure 5.17   PicBasic program of Project 4

PicBasic Pro

The software for PicBasic Pro language is given in Figure 5.18. At the beginning of the program TRISB is cleared to 0 to make all PORTB pins as outputs. Then the hexadecimal number $AA is sent to PORTB to turn on the odd-numbered LEDs.

```
'***********************************************************
'
'
'               TURN ON ODD NUMBERED LEDS
'               ===============================
'
' This program turns on odd numbered LEDs (bit 1, bit 3, bit 5, bit 7) connected
' to PORTB of a PIC16F627 microcontroller.
'
'
'
' Author:          Dogan Ibrahim
' Date:            October, 2005
' Compiler:        PicBasic Pro
' File:            LED6.BAS
'
' Modifications
' ==========
'
'***********************************************************
'
' DEFINITIONS
'
'
' START OF MAIN PROGRAM
'
      TRISB = 0                    ' Set PORTB pins as outputs
      PORTB = $AA                  ' Turn on odd numbered LEDs

      END                          ' End of program
```

Figure 5.18   PicBasic Pro program of Project 5

Project 5

Project title: Binary counting LEDs

Project description: In this project, 8 LEDs are connected to PORTB of a PIC microcontroller. When the project is started (or when reset), the LEDs count in binary with a 250 ms delay between each count as shown in Figure 5.19. The count goes from 0 (binary "00000000") to 255 (binary "11111111") and then repeats forever.

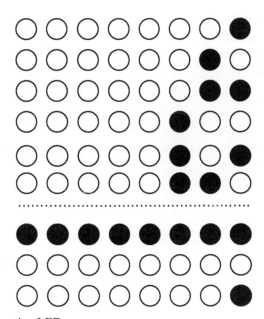

Figure 5.19   Binary counting LEDs

Hardware: The circuit diagram and the construction of the project are as in Figures 5.14 and 5.15, respectively. A PIC16F627 model PIC microcontroller is used and the microcontroller is operated from its 4 MHz internal clock. The LEDs are connected to 8 pins of PORTB using 330 Ω current-limiting resistors. An external reset button is connected to MCLR input of the microcontroller.

Flow diagram: The flow diagram of the project is shown in Figure 5.20. At the beginning of the program the I/O direction is specified. A byte variable called *Cnt* is used as the loop variable and it is incremented by one every 250 ms. When *Cnt* reaches 255 it overflows and takes the next value 0 and this process is repeated forever.

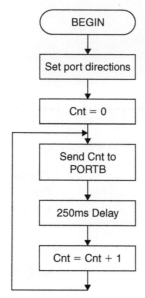

Figure 5.20   Flow diagram of Project 5

Software: **PicBasic**

The software for PicBasic language is given in Figure 5.21. At the beginning of the program PORTB and TRISB addresses are defined. TRISB is then cleared to 0 to make all PORTB pins as outputs. Then variable *Cnt* is initialised to zero. Inside the program loop the value of *Cnt* is sent to PORTB and then incremented by one. This loop is repeated forever.

```
'*****************************************************************
'
'
'                   BINARY COUNTING LEDS
'                   ====================
'
'
' 8 LEDs are connected to PORTB of a PIC microcontroller. This program
' counts in binary and displays the result on the LEDs with 250ms delay
' between each count.
'
'
'
' Author:          Dogan Ibrahim
' Date:            October, 2005
```

Figure 5.21   (Continued)

```
' Compiler:      PicBasic
' File:          LED7.BAS
'
' Modifications
' ==========
'
'*********************************************************************
'
' SYMBOLS
'
Symbol TRISB = $86              ' TRISB address
Symbol PORTB = $06              ' PORTB address
Symbol Cnt = B0                 ' Cnt is a byte variable
'
' START OF MAIN PROGRAM
'
     POKE TRISB, 0              ' Set PORTB pins as outputs

LOOP:
     POKE PORTB, Cnt            ' Send Cnt to PORTB
     PAUSE 250                  ' Wait 250ms
     Cnt = Cnt +1               ' Increment Cnt
     GOTO LOOP                  ' Repeat

     END                        ' End of program
```

Figure 5.21   PicBasic program of Project 5

PicBasic Pro

The software for PicBasic Pro language is given in Figure 5.22. At the beginning of the program TRISB is cleared to 0 to make all PORTB pins as outputs. Then variable *Cnt* is initialised to zero. Inside the program loop the value of *Cnt* is sent to PORTB and then incremented by one. Variable *Cnt* is a byte and increments from 0 to 255 and then overflows back to 0. The loop is repeated forever.

```
'*********************************************************************
'
'              BINARY COUNTING LEDS
'              =====================
'
' 8 LEDs are connected to PORTB of a PIC16F627 microcontroller.
' This program counts in binary and displays the result on the LEDs. A
' 250ms delay is used between each count.
```

Figure 5.22   (Continued)

```
'
'
'
' Author:         Dogan Ibrahim
' Date:           October, 2005
' Compiler:       PicBasic Pro
' File:           LED8.BAS
'
' Modifications
' ==========
'
'*********************************************************************
'
' DEFINITIONS
'
Cnt VAR Byte                          ' Declare Cnt as a Byte variable
'
' START OF MAIN PROGRAM
'
    TRISB = 0                         ' Set PORTB pins as outputs
    Cnt = 0                           ' Initialise Cnt to 0
LOOP:
    PORTB = Cnt                       ' Send Cnt to PORTB
    PAUSE 250                         ' 250ms delay
    Cnt = Cnt + 1                     ' Increment Cnt
    GOTO LOOP                         ' Repeat

    END                               ' End of program
```

Figure 5.22   PicBasic Pro program of Project 5

Project 6

Project title: Left scrolling LEDs

Project description: In this project, 8 LEDs are connected to PORTB of a PIC microcontroller. When the project is started (or when reset), the LEDs scroll to the left with a 250 ms delay between each output as shown in Figure 5.23. When the left-most LED (bit 7) is lit, the next LED lit is the right-most LED (bit 0). This process is repeated forever.

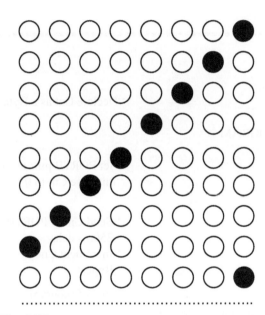

Figure 5.23   Left scrolling LEDs

Hardware: The circuit diagram and the construction of the project are as in Figures 5.14 and 5.15, respectively. A PIC16F627 model PIC microcontroller is used and the microcontroller is operated from its 4 MHz internal clock. The LEDs are connected to 8 pins of PORTB using 330 Ω current-limiting resistors. An external reset button is connected to MCLR input of the microcontroller.

Flow diagram: The flow diagram of the project is shown in Figure 5.24. At the beginning of the program the I/O direction is specified. A byte variable called *Cnt* is used as the loop variable and it is shifted left by one digit at every iteration of the loop. When the value of *Cnt* is 128 (left-most LED is on), it is re-initialised back to 1. A 250 ms delay is used between each output.

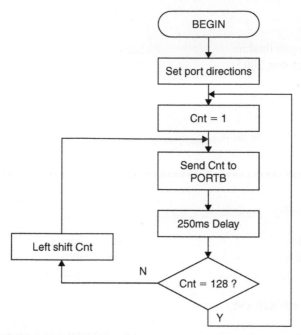

Figure 5.24 Flow diagram of Project 6

Software: **PicBasic**
The software for PicBasic language is given in Figure 5.25. At the beginning of the program PORTB and TRISB addresses are defined. TRISB is then cleared to 0 to make all PORTB pins as outputs. Then variable *Cnt* is initialised to 1 and its value is sent to PORTB to turn on the right-most LED (bit 0). Inside the program loop the value of *Cnt* is shifted left one digit by multiplying with 2 so that the next higher LED can be turned on. When the left-most LED (bit 7) is turned on the value of *Cnt* is 128 and it is re-initialised to 1 so that the next LED to be turned on is the first LED (bit 0). This loop is repeated forever with 250 ms delay between each output.

```
'******************************************************************
'
'
'                LEFT SCROLLING LEDS
'                =====================
'
'
' 8 LEDs are connected to PORTB of a PIC microcontroller. This program
' scrolls the LEDs to the left by one digit. When the LED at bit 7 is turned
' on, then the next LED to be turned on is the LED at bit position 0. The
' program loop is repeated with 250ms delay between each loop.
```
Figure 5.25   (Continued)

```
'
'
' Author:          Dogan Ibrahim
' Date:            October, 2005
' Compiler:        PicBasic
' File:            LED9.BAS
'
' Modifications
' ==========
'
'***************************************************************
'
' SYMBOLS
'
Symbol TRISB = $86                    ' TRISB address
Symbol PORTB = $06                    ' PORTB address
Symbol Cnt = B0                       ' Cnt is a byte variable
'
' START OF MAIN PROGRAM
'
        POKE TRISB, 0                 ' Set PORTB pins as outputs

INIT:
        CNT = 1                       ' Initialise Cnt to 1

LOOP:
        POKE PORTB, Cnt               ' Send Cnt to PORTB
        PAUSE 250                     ' Wait 250ms
        IF Cnt = 128 THEN INIT        ' IF the left-most LED
        Cnt = Cnt * 2                 ' Left-shift Cnt by 1 digit
        GOTO LOOP                     ' Repeat

        END                           ' End of program
```

Figure 5.25   PicBasic program of Project 6

PicBasic Pro

The software for PicBasic Pro language is given in Figure 5.26. At the beginning of the program TRISB is cleared to 0 to make all PORTB pins as outputs. Then variable *Cnt* is initialised to 1 and its value is sent to PORTB to turn on the right-most LED (LED at bit position 0). Inside the program loop the value of *Cnt* is shifted left one digit by using the shift operator "$\ll$" so that the next higher LED can be turned on. When the left-most LED (bit 7) is turned on the value of *Cnt* is 128 and it is re-initialised to 1 for the next loop. This loop is repeated forever with a 250 ms delay between each output.

```
'**********************************************************************
'
'                    LEFT SHIFTING LEDS
'                    ==================
'
' 8 LEDs are connected to PORTB of a PIC16F627 microcontroller.
' This program scrolls the LEDs left with 250ms delay between each
' output. When the LED at bit 7 is on, the next LED to be on is the
' one at bit position 0.
'
'
' Author:         Dogan Ibrahim
' Date:           October, 2005
' Compiler:       PicBasic Pro
' File:           LED10.BAS
'
' Modifications
' =============
'
'**********************************************************************
'
' DEFINITIONS
'
Cnt VAR Byte                            ' Declare Cnt as a Byte variable
'
' START OF MAIN PROGRAM
'
      TRISB = 0                         ' Set PORTB pins as outputs
INIT:
       Cnt = 1                          ' Initialise Cnt to 1
LOOP:
       PORTB = Cnt                      ' Send Cnt to PORTB
       PAUSE 250                        ' Wait 250ms
       IF Cnt = 128 THEN INIT           ' If the left-most LED
       Cnt = Cnt << 1                   ' Left-shift Cnt by 1 digit
       GOTO LOOP                        ' Repeat

       END                             ' End of program
```

Figure 5.26   PicBasic Pro program of Project 6

Project 7

Project title: Right scrolling LEDs

Project description: In this project, 8 LEDs are connected to PORTB of a PIC microcontroller. When the project is started (or when reset), the LEDs scroll to the right with a 250 ms delay between each output as shown in Figure 5.27. When the right-most LED (bit 0) is lit, the next LED lit is the left-most LED (bit 7). This process is repeated forever.

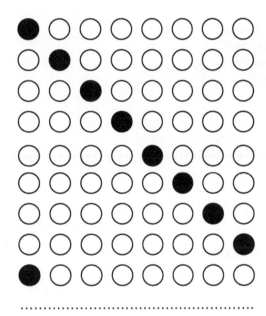

Figure 5.27   Right scrolling LEDs

Hardware: The circuit diagram and the construction of the project are as in Figures 5.14 and 5.15, respectively. A PIC16F627 model PIC microcontroller is used and the microcontroller is operated from its 4 MHz internal clock. The LEDs are connected to 8 pins of PORTB using 330 Ω current-limiting resistors. An external reset button is connected to MCLR input of the microcontroller.

Flow diagram: The flow diagram of the project is shown in Figure 5.28. At the beginning of the program the I/O direction is specified. A byte variable called *Cnt* is used as the loop variable and it is shifted right by one digit at every iteration of the loop. When the value of *Cnt* is 1 (the right-most LED is on), it is re-initialised back to 128. A 250 ms delay is used between each output.

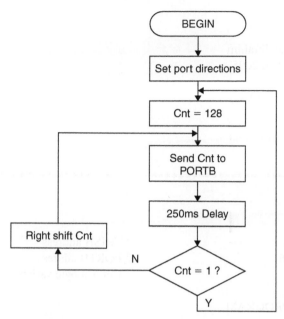

Figure 5.28   Flow diagram of Project 7

Software: **PicBasic**
The software for PicBasic language is given in Figure 5.29. At the beginning
of the program PORTB and TRISB addresses are defined. TRISB is then
cleared to 0 to make all PORTB pins as outputs. Then variable *Cnt* is initialised to 128 and its value is sent to PORTB to turn on the left-most LED
(bit 7). Inside the program loop the value of *Cnt* is shifted right one digit by
dividing with 2 so that the next lower LED can be turned on. When the right-
most LED (bit 0) is turned on the value of *Cnt* is 1 and it is re-initialised to
128 so that the next LED to be turned on is the left-most LED (bit 7). This
loop is repeated forever with 250 ms delay between each output.

```
'*******************************************************************
'
'
'           RIGHT SCROLLING LEDS
'           ==========================
'
'
' 8 LEDs are connected to PORTB of a PIC microcontroller. This program
' scrolls the LEDs to the right by one digit. When the LED at bit 0 is turned
' on, then the next LED to be turned on is the LED at bit position 7. The
' program loop is repeated with 250ms delay between each loop.
```

Figure 5.29   (Continued)

```
'
'
' Author:        Dogan Ibrahim
' Date:          October, 2005
' Compiler:      PicBasic
' File:          LED11.BAS
'
' Modifications
' ==========
'
'***********************************************************************
'
' SYMBOLS
'
Symbol TRISB = $86                    ' TRISB address
Symbol PORTB = $06                    ' PORTB address
Symbol Cnt = B0                       ' Cnt is a byte variable
'
' START OF MAIN PROGRAM
'
     POKE TRISB, 0                    ' Set PORTB pins as outputs
INIT:
     CNT = 128                        ' Initialise Cnt to 128

LOOP:
     POKE PORTB, Cnt                  ' Send Cnt to PORTB
     PAUSE 250                        ' Wait 250ms
     IF Cnt = 1 THEN INIT             ' IF the right-most LED
     Cnt = Cnt / 2                    ' Right-shift Cnt by 1 digit
     GOTO LOOP                        ' Repeat

     END                             ' End of program
```

Figure 5.29   PicBasic program of Project 7

PicBasic Pro

The software for PicBasic Pro language is given in Figure 5.30. At the beginning of the program TRISB is cleared to 0 to make all PORTB pins as outputs. Then variable *Cnt* is initialised to 128 and its value is sent to PORTB to turn on the left-most LED (LED at bit position 7). Inside the program loop the value of *Cnt* is shifted right one digit by using the shift operator ">>" so that the next lower LED can be turned on. When the right-most LED (bit 0) is turned on the value of *Cnt* is 1 and it is re-initialised to 128 for the next loop. This loop is repeated forever with a 250 ms delay between each output.

```
'*********************************************************************
'
'                    RIGHT SHIFTING LEDS
'                    ===================
'
' 8 LEDs are connected to PORTB of a PIC16F627 microcontroller.
' This program scrolls the LEDs right with 250ms delay between each
' output. When the LED at bit 0 is on, the next LED to be on is the
' one at bit position 7.
'
'
' Author:        Dogan Ibrahim
' Date:          October, 2005
' Compiler:      PicBasic Pro
' File:          LED12.BAS
'
' Modifications
' ===========
'
'*********************************************************************
'
' DEFINITIONS
'
Cnt VAR Byte                          ' Declare Cnt as a Byte variable
'
' START OF MAIN PROGRAM
'
      TRISB = 0                       ' Set PORTB pins as outputs
INIT:
      Cnt = 128                       ' Initialise Cnt to 128
LOOP:
      PORTB = Cnt                     ' Send Cnt to PORTB
      PAUSE 250                       ' Wait 250ms
      IF Cnt = 1 THEN INIT            ' If the right-most LED
      Cnt = Cnt >> 1                  ' Right-shift Cnt by 1 digit
      GOTO LOOP                       ' Repeat

      END                             ' End of program
```

Figure 5.30   PicBasic Pro program of Project 7

Project 8

Project title: Right-left scrolling LEDs

Project description: In this project, 8 LEDs are connected to PORTB of a PIC microcontroller. Also a push-button switch is connected to bit 0 of PORTA using a pull-up resistor. Normally the LEDs scroll to the left as in Project 6. When the switch is pressed the LEDs scroll to the right as in Project 7.

Hardware: The circuit diagram of the project is shown in Figure 5.31. The circuit is very similar to Figure 5.14, but in this project additionally a switch is connected to bit 0 of PORTA to control the direction of scrolling. A PIC16F627 model PIC microcontroller is used and the microcontroller is operated from its 4 MHz internal clock. The LEDs are connected to 8 pins of PORTB using 330 Ω current-limiting resistors. An external reset button is connected to MCLR input of the microcontroller.

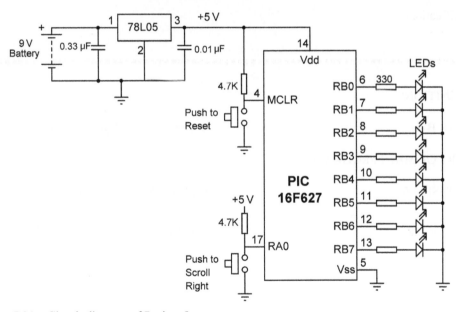

Figure 5.31   Circuit diagram of Project 8

Flow diagram: The flow diagram of the project is shown in Figure 5.32. At the beginning of the program the I/O direction is specified. A byte variable called *Cnt* is used as the loop variable. The program consists of an indefinite loop and at the beginning of the loop the switch is tested. If the switch is logic 1 (i.e. switch is not pressed) then the scrolling is to the left and if the switch is pressed the switch is at logic 0 and scrolling is to the right. A 250 ms delay is used between each output.

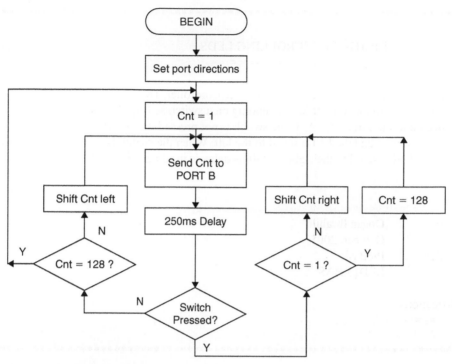

Figure 5.32 Flow diagram of Project 8

Software: **PicBasic**
The software for PicBasic language is given in Figure 5.33. At the beginning of the program PORTA, PORTB, TRISA, TRISB, and CMCON register addresses are defined. TRISA is set to 1 so that bit 0 of PORTA is configured as an input port. Similarly, TRISB is cleared to 0 so that all bits of PORTB are configured as outputs. Push-button switch is connected to bit 0 of PORTA (RA0). Normally this pin is pulled high to logic 1 by using a resistor. When the switch is pressed the pin goes down to logic 0. PORTA pins on the PIC16F627 microcontroller have dual functions and they can either be used as analog comparator inputs, or as digital I/O ports. CMCON register is used to control the function of these pins. Setting CMCON to 7 configures PORTA pins as digital I/O ports.

Inside the LOOP, the value of *Cnt* is sent to PORTB and the PEEK instruction is used to read the switch setting. "Bit0" refers to bit 0 of variable "B0" which is where the switch is connected. When the switch is pressed the program jumps to label PRESSED where the LEDs are scrolled right. When the switch is not pressed the LEDs are scrolled left. This loop is repeated forever with 250 ms delay between each output.

```
'***********************************************************************
'
'                   RIGHT-LEFT SCROLLING LEDS
'                   ===========================
'
' 8 LEDs are connected to PORTB of a PIC microcontroller. This program
' scrolls the LEDs to the right or left depending on a switch setting. The switch
' is connected to bit 0 of PORT A. If the switch is not pressed the switch
' output is at logic 1 and the LEDs scroll to the left. When the switch is
' pressed the LEDs scroll to the right. A 250ms delay is used between each
' output.
'
'
' Author:        Dogan Ibrahim
' Date:          October, 2005
' Compiler:      PicBasic
' File:          LED13.BAS
'
' Modifications
' ============
'
'***********************************************************************
'
' SYMBOLS
'
Symbol TRISA = $85                    ' TRISA address
Symbol TRISB = $86                    ' TRISB address
Symbol PORTA = $05                    ' PORTA address
Symbol PORTB = $06                    ' PORTB address
Symbol CMCON = $1F                    ' CMCON address
Symbol Cnt = B1                       ' Cnt is a byte variable
Symbol Switch = B0                    ' Switch is a byte variable
'
' START OF MAIN PROGRAM
'
      POKE CMCON, 7                   ' RA0-RA3 are digital I/O
      POKE TRISA, 1                   ' Set PORTA bit 0 as input
      POKE TRISB, 0                   ' Set all PORTB pins as outputs
INIT:
      CNT = 1                         ' Initialise Cnt to 1
```

Figure 5.33   (Continued)

```
LOOP:
        POKE PORTB, Cnt                    ' Send Cnt to PORTB
        PAUSE 250                          ' Wait 250ms
        PEEK PORTA, Switch                 ' Read switch setting
        IF Bit0 = 0 THEN PRESSED           ' If switch is pressed
        IF Cnt = 128 THEN INIT
        Cnt = Cnt * 2                      ' Shift Cnt left
        GOTO LOOP

PRESSED:                                   ' Switch is pressed
        IF Cnt = 1 THEN NXT
        Cnt = Cnt / 2
        GOTO LOOP

NXT:
        Cnt = 128
        GOTO LOOP

        END                                ' End of program
```

Figure 5.33   PicBasic program of Project 8

PicBasic Pro

The software for PicBasic Pro language is given in Figure 5.34. The PicBasic Pro program is easier to understand. At the beginning of the program TRISB is cleared to 0 to make all PORTB pins as outputs. Also, TRISA is set to 1 so that bit 0 of PORTA is configured as input. CMCON register is set to 7 to configure PORTA pins as digital I/O.

The switch setting is then checked using an IF statement. When the switch is pressed bit 0 of PORTA goes to logic 0 and the program scrolls the LEDs to right. When the switch is not pressed bit 0 of PORTA is at logic 1 and the program scrolls the LEDs to the left.

```
'******************************************************************
'
'                   RIGHT-LEFT SHIFTING LEDS
'                   =========================
'
'
' 8 LEDs are connected to PORTB of a PIC16F627 microcontroller.
' This program scrolls the LEDs right or left depending on the mode of a
' push-button switch. When the switch is not pressed LEDs are scrolled left.
' When the switch is pressed, LEDs are scrolled right. A 250ms delay
' is used between each output.
```

Figure 5.34   (Continued)

```
'
'
' Author:        Dogan Ibrahim
' Date:          October, 2005
' Compiler:      PicBasic Pro
' File:          LED14.BAS
'
' Modifications
' ==========
'
'*************************************************************************
'
' DEFINITIONS
'
Cnt VAR Byte                            ' Declare Cnt as a Byte variable
'
' START OF MAIN PROGRAM
'
        CMCON = 7                       ' Set PORTA as digital I/O
        TRISA = 1                       ' Set RA0 as input
        TRISB = 0                       ' Set PORTB pins as outputs
INIT:
        Cnt = 1                         ' Initialise Cnt to 1
LOOP:
        PORTB = Cnt                     ' Send Cnt to PORTB
        PAUSE 250                       ' Wait 250ms
        IF PORTA.0 = 0 THEN
           IF Cnt = 1 THEN Cnt = 128: GOTO LOOP
           Cnt = Cnt >> 1               ' Shift right
           GOTO LOOP
        ELSE
           IF Cnt = 128 THEN INIT
           Cnt = Cnt << 1               ' Shift left
           GOTO LOOP
        ENDIF

        END                             ' End of program
```

Figure 5.34   PicBasic Pro program of Project 8

Project 9

Project title: LED dice

Project description: In this project, 7 LEDs are connected to PORTB of a PIC microcontroller and arranged such that they can show the faces of a dice when lit. Also a push-button switch is connected to bit 7 of PORTB using a pull-up resistor. When the switch is pressed the LEDs are lit randomly to show a dice number between 1 and 6. Figure 5.35 shows the LEDs lit for a given dice number.

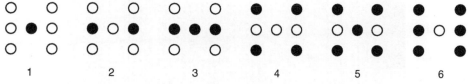

Figure 5.35   LED dice

Hardware: The circuit diagram of the project is shown in Figure 5.36. The 7 LEDs are connected to bit 0 to bit 6 of PORTB. Bit 7 of PORTB is connected to a push-button switch which simulates the throwing of a dice when pressed.

                     A PIC16F627 model PIC microcontroller is used and the microcontroller is operated from its 4 MHz internal clock. The LEDs are connected to

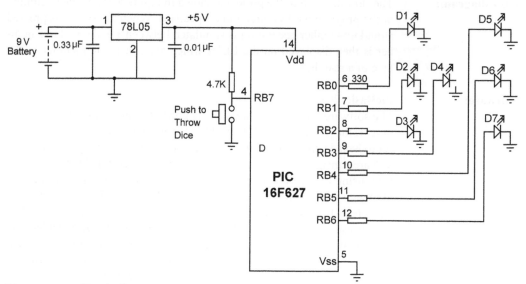

Figure 5.36   Circuit diagram of Project 9

8 pins of PORTB using 330 Ω current-limiting resistors. External reset button is not used and power-on-reset is used with the MCLR disabled during the programming of the device.

The project constructed on a breadboard is shown in Figure 5.37.

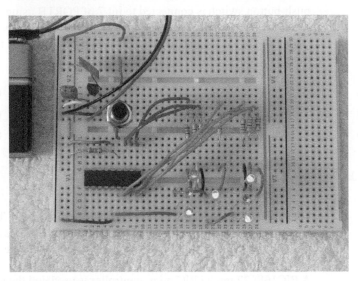

Figure 5.37 Construction of the project

Flow diagram: The flow diagram of the project is shown in Figure 5.38. At the beginning of the program the I/O direction is specified. When the switch is pressed a random number generation is simulated between 1 and 6 and this number is then displayed on the LEDs which are constructed similar to the face of a real dice.

Software: **PicBasic**
The software for PicBasic language is given in Figure 5.39. At the beginning of the program PORTB and TRISB addresses are defined. TRISB is set to hexadecimal $80 so that bit 7 is configured as input port and bits 0 to 6 are configured as output ports. A number is generated between 1 and 6 by using a loop. Inside this loop if the switch is not pressed the dice number is incremented by one between 1 and 6. When the number reaches 7 it is reset back to 1 so that the generated number is between 1 and 6. The loop is executed so fast that the generated numbers can be considered to be random. When the switch is pressed, the program jumps to label NEWNO and here the current number (in variable *DICE*) is used in a LOOKUP statement to determine the LEDs to be turned on. If the value of *DICE* is 1, variable LEDS is loaded with $08. Similarly, if the value of

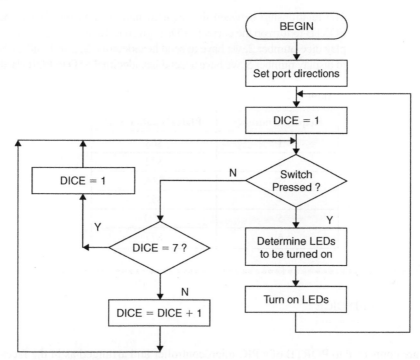

Figure 5.38   Flow diagram of Project 9

DICE is 2, variable LEDS is loaded with $22 and so on. The required LEDs are then turned on to display the number similar to the dots on a dice. Table 5.1 shows the relationship between the number to be displayed and the LEDs to be turned on to display this number. For example, to display number 1 (i.e. only the middle LED is on), we have to turn on D4. Similarly, to display number 4, we have to turn on D1, D3, D5, and D7.

Table 5.1   Required number and LED to be turned on

Required number	LEDs to be turned on
1	D4
2	D2, D6
3	D2, D4, D6
4	D1, D3, D5, D7
5	D1, D3, D4, D5, D7
6	D1, D2, D3, D5, D6, D7

The relationship between the required number and the data to be sent to PORTB to turn on the correct LEDs is given in Table 5.2. For example, to display dice number 2, we have to send hexadecimal $22 to PORTB. Similarly, to display number 5, we have to send hexadecimal $5D to PORTB and so on.

Table 5.2   Required number and PORTB data

Required number	PORTB data (Hex)
1	$08
2	$22
3	$2A
4	$55
5	$5D
6	$77

```
'********************************************************************
'
'                LED DICE
'                ========
'
' 7 LEDs are connected to PORTB of a PIC microcontroller and arranged as in the faces
' of a dice. Also, a push-button switch is connected to bit 7 of PORTB.
' When the switch is pressed the program generates a dice number
' between 1 and 6 and turns on the appropriate LEDs to imitate the faces
' of a real dice. The LEDs are turned on for 5 seconds and after this time
' they are cleared and the program is ready to accept a new push-button
' action.
'
' The microcontroller is operated with internal 4MHz clock and internal
' power-on-reset.
'
'
'
' Author:        Dogan Ibrahim
' Date:          October, 2005
' Compiler:      PicBasic
' File:          LED15.BAS
'
' Modifications
' =============
'
'********************************************************************
```

Figure 5.39   (Continued)

```
'
' SYMBOLS
'
Symbol TRISB = $86          ' TRISB address
Symbol PORTB = $06          ' PORTB address
Symbol Switch = B0          ' Switch is a byte variable
Symbol LEDS = B1            ' LEDs to be turned on
Symbol DICE = B2            ' Dice number (between 1 and 6)
'
' START OF MAIN PROGRAM
'
     POKE TRISB, $80        ' RB7 input, RB0-RB6 outputs
'
' Wait until switch is pressed
'
WAIT:
     DICE = 1
NXT: PEEK PORTB, Switch     ' Check if switch is pressed
     IF Bit7 = 0 THEN NEWNO ' If pressed goto NEWNO
     DICE = DICE + 1        ' Increment dice number
     IF DICE = 7 THEN WAIT  ' between 1 and 6
     GOTO NXT               ' repeat
NEWNO:
'
' Find the LEDs to be turned on. DICE is between 1 and 6. LEDS is the data
' to be sent to PORTB to turn on the required LEDs
'
     LOOKUP DICE, (0, $08, $22, $2A, $55, $5D, $77), LEDS
'
' Turn on the LEDs
'
     POKE PORTB, LEDS

     PAUSE 5000             ' Wait 5 seconds
     POKE PORTB, 0          ' Turn off all LEDs
     GOTO WAIT              ' Repeat

     END                    ' End of program
```

Figure 5.39   PicBasic program of Project 9

PicBasic Pro

The software for PicBasic Pro language is given in Figure 5.40. At the beginning of the program TRISB is set to $80 to configure bit 7 of PORTB as input and the other PORTB pins as outputs. The switch is then checked inside a loop using an IF statement. If the switch is not pressed variable DICE is incremented by one between 1 and 6. When the switch is pressed the current value of *DICE* is taken and used in a LOOKUP statement to obtain the data to be sent to PORTB so that the correct LEDs can be turned on.

```
'********************************************************************
'
'                    LED DICE
'                    ========
'
' 7 LEDs are connected to PORTB of a PIC microcontroller and arranged as in the faces
' of a dice. Also, a push-button switch is connected to bit 7 of PORTB.
' When the switch is pressed the program generates a dice number
' between 1 and 6 and turns on the appropriate LEDs to imitate the faces
' of a real dice. The LEDs are turned on for 5 seconds and after this time
' they are cleared and the program is ready to accept a new push-button
' action.
'
' The microcontroller is operated with internal 4MHz clock and internal
' power-on-reset.
'
'
'
' Author:       Dogan Ibrahim
' Date:         October, 2005
' Compiler:     PicBasic Pro
' File:         LED16.BAS
'
' Modifications
' =============
'
'********************************************************************
'
' DEFINITIONS
'
LEDS    VAR     BYTE
DICE    VAR     BYTE
'
' START OF MAIN PROGRAM
'
      TRISB = $80                          ' RB7 input, RB0-RB6 outputs
```

Figure 5.40   (Continued)

```
'
' Wait until switch is pressed
'
AGAIN:
    DICE = 1
NXT: IF PORTB.7 = 0 THEN NEWNO          ' If pressed goto NEWNO
    DICE = DICE + 1                     ' Increment dice number
    IF DICE = 7 THEN AGAIN              ' between 1 and 6
    GOTO NXT                            ' repeat
NEWNO:
'
' Find the LEDs to be turned on. DICE is between 1 and 6. LEDS is the data
' to be sent to PORTB to turn on the required LEDs
'
    LOOKUP DICE, [0, $08, $22, $2A, $55, $5D, $77], LEDS
'
' Turn on the LEDs
'
    PORTB = LEDS                        ' Turn on appropriate LEDs

    PAUSE 5000                          ' Wait 5 seconds
    PORTB = 0                           ' Turn off all LEDs
    GOTO AGAIN                          ' Repeat

    END                                 ' End of program
```

Figure 5.40   PicBasic Pro program of Project 9

Project 10

Project title: 7-segment LED display counter

Project description: In this project, a 7-segment LED display is used as a counter. Numbers
from 0 to 9 are displayed on the display continuously as 0 1 2 3 ... 8 9 0
1 2 ... with 1 s delay between each count.

Hardware: 7-Segment displays are frequently used in electronic circuits as indicators.
As shown in Figure 5.41, a 7-segment display basically consists of 7 LEDs
connected such that numbers from 0 to 9 and some letters can be displayed.
Figure 5.42 shows the segment names of a typical 7-segment display.

Figure 5.41   Some 7-segment displays

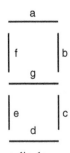

Figure 5.42   Segment names of a 7-segment display

Figure 5.43 shows how numbers from 0 to 9 can be obtained by turning on different segments of the display.

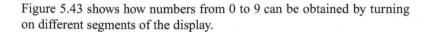
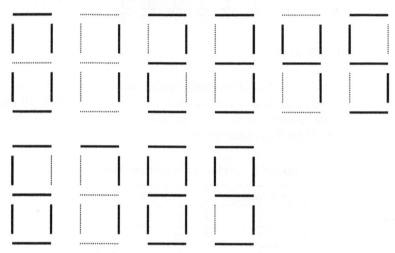

Figure 5.43 Obtaining numbers 0–9

7-segment displays are available in two different configurations: *common cathode* and *common anode*. As shown in Figure 5.44, in common cathode displays the cathodes of all the segment LEDs are tied together and then this common point is connected to ground. A required segment is then turned on by applying a logic 1 to the anode of this segment. Here, the output pin of the microcontroller is in current sourcing mode.

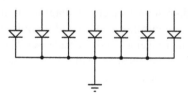

Figure 5.44 Common cathode display

In a common-anode display, the anodes of all the segment LEDs are tied together (see Figure 5.45) and then this common point is connected to V supply voltage. A required segment is turned on by applying a logic 0 to the cathode of this segment. Here, the output pin of the microcontroller is in current sinking mode.

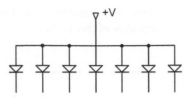

Figure 5.45   Common-anode display

In this project, a *Kingbright SA52-11* model red common anode display is used. This is a 13 mm height (0.52 inch) display with 10 pins. The pin configuration is as shown in Table 5.3. The display also has a segment LED for the decimal point.

Table 5.3   SA52-11 pin configuration

Pin Number	Segment
1	E
2	D
3	Common anode
4	C
5	Decimal point
6	B
7	A
8	Common anode
9	F
10	g

Figure 5.46 shows the circuit diagram of the project. A PIC16F627 model PIC microcontroller is used and the microcontroller is operated from its 4 MHz internal clock and internal reset. The display is connected to PORTB of the microcontroller using 330 Ω current-limiting resistors in each segment of the display.

The project constructed on a breadboard is shown in Figure 5.47.

The relationship between the displayed numbers and the data to be sent to PORTB is shown in Table 5.4. The display is connected to the microcontroller using segments a to g. In this Table, x is a don'tcare entry, taken as 0 and is used to make the bit number 8. For example, to display number 2, we have to send hexadecimal number $5B to PORTB. Similarly, to display number 8, we have to send hexadecimal number $7F to PORTB.

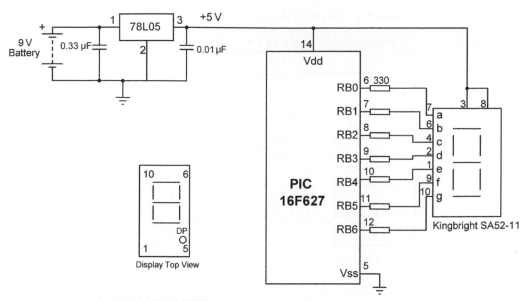

Figure 5.46   Circuit diagram of Project 10

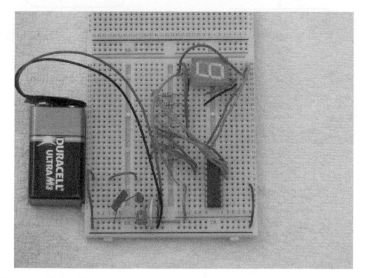

Figure 5.47   Construction of the project

Flow diagram: The flow diagram of the project is shown in Figure 5.48. At the beginning of the program the I/O direction is specified by loading 0 to TRISB, i.e. all PORTB pins are configured as output pins. Then a loop is formed to send numbers 0 to 9 to the display. Inside the loop subroutine CONVERT

Table 5.4   Displayed number and data sent to PORTB

Number	x g f e d c b a	PORTB data
0	0 0 1 1 1 1 1 1	$3F
1	0 0 0 0 0 1 1 0	$06
2	0 1 0 1 1 0 1 1	$5B
3	0 1 0 0 1 1 1 1	$4F
4	0 1 1 0 0 1 1 0	$66
5	0 1 1 0 1 1 0 1	$6D
6	0 1 1 1 1 1 0 1	$7D
7	0 0 0 0 0 1 1 1	$07
8	0 1 1 1 1 1 1 1	$7F
9	0 1 1 0 1 1 1 1	$6F

x is not used, taken as 0.

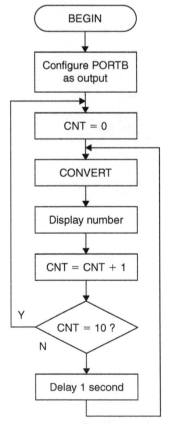

Subroutine CONVERT

Figure 5.48   Flow diagram of Project 10

is called to determine the actual data to be sent to PORTB in order to display the required number. This subroutine uses the LOOKUP statement to determine the bit segments to be turned on for a required number. The data to be sent to PORTB is inverted since we are using a common-anode display (i.e. a segment is turned on by making the segment pin logic 0). The process is repeated after a 1 s delay between each output.

Software: **PicBasic**

The software for PicBasic language is given in Figure 5.49. At the beginning of the program PORTB and TRISB addresses are defined. Also, variables *Cnt* and *Pattern* are declared as byte variables. TRISB is cleared to 0 so that PORTB pins are configured as outputs. At the beginning of the program variable *Cnt* is cleared to 0 and subroutine CONVERT is called. This subroutine receives *Cnt* as the input variable and returns the bit pattern in variable *Pattern*. For example, if *Cnt* is 0, *Pattern* is assigned hexadecimal number $3F, if *Cnt* is 1, *Pattern* is assigned $06, etc. The bit pattern is then inverted since we are using a common-anode type display (a segment is turned on by clearing the segment pin). Variable *Pattern* is inverted by performing a bit-wise *Exclusive OR* with hexadecimal number $FF (a bit is inverted when it is exclusive or'ed with 1). Variable *Cnt* is then incremented and cleared to 0 when it reaches 10 so that the number is between 0 and 9. Otherwise, the program jumps to label NXT where the next number is displayed. After displaying a number, the program waits for 1 s and the process repeats forever.

```
'******************************************************************
'
'                7-SEGMENT DISPLAY COUNTER
'                =============================
'
'
' In this project a common-anode type 7-segment display is connected
' to PORTB of a PIC16F627 model microcontroller. The project displays
' numbers 0 to 9 on the display with 1 second delay between each output.
' The microcontroller is operated with the internal 4MHz clock and also
' the internal reset is used.
'
' The connection between the microcontroller and the display is as
' follows:
'
'        RB0      segment a
'        RB1      segment b
```

Figure 5.49   (Continued)

```
'      RB2      segment c
'      RB3      segment d
'      RB4      segment e
'      RB5      segment f
'      RB6      segment g
'
' The decimal point of the display is not used.
'
'
' Author:        Dogan Ibrahim
' Date:          October, 2005
' Compiler:      PicBasic
' File:          LED17.BAS

' Modifications
' ============
'
'*********************************************************************
'
' SYMBOLS
'
Symbol TRISB = $86                      ' TRISB address
Symbol PORTB = $06                      ' PORTB address
Symbol Cnt = B0                         ' Cnt is a byte variable
Symbol Pattern = B1                     ' Pattern is a byte variable
'
' START OF MAIN PROGRAM
'
      POKE TRISB, 0                     ' PORTB is output

LOOP:
      Cnt = 0                           ' Initialise CNT to 0
NXT:
      GOSUB CONVERT                     ' Find the bit pattern to send to PORTB
      POKE PORTB, Pattern               ' Send Pattern to PORTB
      Cnt = Cnt + 1                     ' Increment count
      PAUSE 1000                        ' Wait 1 second
      IF CNT = 10 THEN LOOP
      GOTO NXT
```

Figure 5.49   (Continued)

CONVERT:
'
' Find the bit pattern to be sent to PORTB in order to turn on the correct segments
' to display the required number. Cnt contains a number between 0 and 9 and
' on return from LOOKUP statement, Pattern contains the bit pattern to send to
' PORTB to display the required number in Cnt. Because we are using a
' common-anode display, a segment is turned on when it is logic 0 and thus the
' bit pattern is inverted before sending to PORTB.
'

```
        LOOKUP Cnt, ($3F, $06, $5B, $4F, $66, $6D, $7D, $07, $7F, $6F), Pattern
        Pattern = Pattern ^ $FF            ' Invert bits of variable Pattern
        RETURN

        END                                ' End of program
```

Figure 5.49   PicBasic program of Project 10

PicBasic Pro
The software for PicBasic Pro language is given in Figure 5.50. At the beginning of the program TRISB is set to 0 to configure PORTB pins as outputs. Variable *Cnt* is then cleared to 0 and subroutine CONVERT is called to find the bit pattern to be displayed. Statement LOOKUP receives *Cnt* as the input variable and returns the required bit pattern in variable *Pattern*. The bit pattern is then inverted and sent to PORTB to turn on the required display segments. Variable *Cnt* is incremented and cleared to 0 when it reaches 10 so that the number is between 0 and 9. The value of *Cnt* is sent to the display every second.

```
'**********************************************************************
'
'                  7-SEGMENT DISPLAY COUNTER
'                  =============================
'
'
```

' In this project a common-anode type 7-segment display is connected
' to PORTB of a PIC16F627 model microcontroller. The project displays
' numbers 0 to 9 on the display with 1 second delay between each output.
' The microcontroller is operated with the internal 4MHz clock and also
' the internal reset is used.

Figure 5.50   (Continued)

```
'
' The connection between the microcontroller and the display is as
' follows:
'
'        RB0      segment a
'        RB1      segment b
'        RB2      segment c
'        RB3      segment d
'        RB4      segment e
'        RB5      segment f
'        RB6      segment g
'
' The decimal point of the display is not used.
'
'
' Author:        Dogan Ibrahim
' Date:          October, 2005
' Compiler:      PicBasic Pro
' File:          LED18.BAS
'
' Modifications
' ===========
'
'************************************************************************
'
' DEFINITIONS
'
Cnt      VAR      Byte
Pattern  VAR      Byte
'
' START OF MAIN PROGRAM
'
        TRISB = 0                        ' PORTB is output

LOOP:
        Cnt = 0                          ' Initialise CNT to 0
NXT: GOSUB CONVERT                       ' Find the bit pattern to send to PORTB
        PORTB = Pattern                  ' Send Pattern to PORTB
        Cnt = Cnt + 1                    ' Increment count
        PAUSE 1000                       ' Wait 1 second
        IF CNT = 10 THEN LOOP
        GOTO NXT
```

Figure 5.50   (Continued)

CONVERT:
'

' Find the bit pattern to be sent to PORTB in order to turn on the correct segments
' to display the required number. Cnt contains a number between 0 and 9 and
' on return from LOOKUP statement, Pattern contains the bit pattern to send to
' PORTB to display the required number in Cnt. Because we are using a
' common-anode display, a segment is turned on when it is logic 0 and thus the
' bit pattern is inverted before sending to PORTB.
'

        LOOKUP Cnt, [$3F, $06, $5B, $4F, $66, $6D, $7D, $07, $7F, $6F], Pattern
        Pattern = Pattern ^ $FF   ' Invert bits of variable Pattern
        RETURN

        END   ' End of program

Figure 5.50   PicBasic Pro program of Project 10

Project 11

Project title: 7-segment LED dice

Project description: In this project, a 7-segment LED display is used as a dice. Normally the display shows a "0" to indicate that it is waiting for a key press. When the external push-button switch is pressed, a dice number is displayed between 1 and 6 for 3 s. After this time the display clears back to "0" to indicate that it is waiting again for a key press.

Hardware: The circuit diagram of this project is similar to Figure 5.46. A common-anode type 7-segment display is connected as in Figure 5.46, and in addition a push-button switch is connected to bit 7 of PORTB. As shown in Figure 5.51, the switch is normally held at logic 1 using a pull-up resistor.

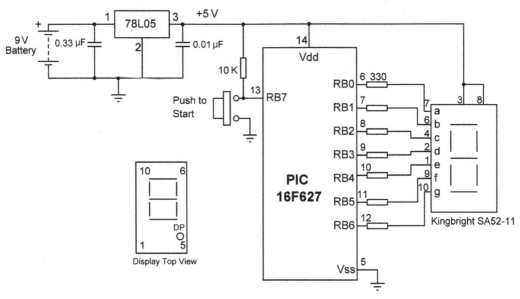

Figure 5.51   Circuit diagram of Project 11

The project constructed on a breadboard is shown in Figure 5.52.

Flow diagram: The flow diagram of the project is shown in Figure 5.53. At the beginning of the program the I/O direction is specified by loading hexadecimal $80 to TRISB, i.e. PORTB pins 0 to 6 are outputs and bit 7 is input. The program then waits for the switch to be pressed. When the switch is pressed, a random number is generated between 1 and 65,535 using the PicBasic

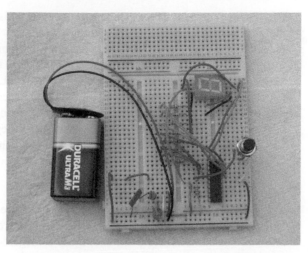

Figure 5.52 Construction of the project

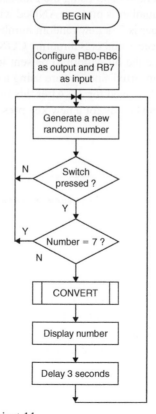

Figure 5.53 Flow diagram of Project 11

RANDOM statement. The generated number is bit-wise ANDed with 7 so that it is between 1 and 7. If the number is 7, a new random number is obtained such that the number is between 1 and 6. Subroutine CONVERT is called to find the bit pattern to be sent to PORTB to turn on the required segments. As in Project 10, this subroutine uses the LOOKUP statement to determine the bit segments to be turned on for a required number. The data to be sent to PORTB is inverted since we are using a common-anode display (i.e. a segment is turned on by making the segment pin logic 0). The dice number is displayed for 3 s. After this time the display is cleared to 0 and the program is ready for a new key press.

Software:

PicBasic

The software for PicBasic language is given in Figure 5.54. At the beginning of the program PORTB and TRISB addresses are defined. TRISB is set to $80 so that bits 0 to 6 of PORTB are configured as outputs and bit 7 is configured as input. The program then waits for the push-button to be pressed. When the key is pressed a new random number is generated between 1 and 65,535 using the PicBasic Pro RANDOM statement. The generated number is bit-wise ANDed with 7 so that it is between 1 and 7. If the number is 7, a new random number is obtained such that the number is between 1 and 6. Subroutine CONVERT uses statement LOOKUP to determine the bit pattern to be sent to PORTB. The data to be sent to PORTB is inverted since we are using a common-anode display. The dice number is displayed for 3 s. After this time the display is cleared to 0 and the program is ready for a new key press. or'ed with 1.

```
'********************************************************************
'
'                  7-SEGMENT DICE
'                  ===============
'
' In this project a common-anode type 7-segment display is connected
' to PORTB of a PIC16F627 model microcontroller. Additionally, a
' push-button switch is connected to bit 7 of PORTB. When the button
' is pressed, the project displays a number between 1 and 6 just like a
' dice. The RANDOM statement is used to generate a random number.
'
' The microcontroller is operated with the internal 4MHz clock and also
' the internal reset is used.
```

Figure 5.54   (Continued)

```
'
' The connection between the microcontrolelr and the display is as
' follows:
'
'         RB0     segment a
'         RB1     segment b
'         RB2     segment c
'         RB3     segment d
'         RB4     segment e
'         RB5     segment f
'         RB6     segment g
'
' The decimal point of the display is not used.
'
'
' Author:        Dogan Ibrahim
' Date:          October, 2005
' Compiler:      PicBasic
' File:          LED19.BAS
'
' Modifications
' ===========
'
'
'*********************************************************************
'
' SYMBOLS
'
Symbol TRISB = $86            ' TRISB address
Symbol PORTB = $06            ' PORTB address
Symbol Switch = B0            ' Switch is a word variable
Symbol Pattern = B1           ' Pattern is a byte variable
Symbol Dice = W1              ' Dice is a word variable
'
' START OF MAIN PROGRAM
'
        POKE TRISB, $80       ' Bits 0-6 are outputs, bit 7 is input

LOOP:                         ' Display 0 at the beginning
        DICE = 0
        GOSUB CONVERT
        POKE PORTB, Pattern   ' Display 0 to show that we are ready
WT:     RANDOM Dice           ' Generate a random number between 0 and 65535
        PEEK PORTB, Switch
        IF Bit7 = 1 THEN WT   ' Wait until switch is pressed
```

Figure 5.54   (Continued)

```
BR:    Dice = Dice & 7              ' Number between 0 and 7
       IF Dice <> 7 THEN NXT       ' If the number is 0 or 7, get a new number
       RANDOM Dice
       GOTO BR
NXT:   GOSUB CONVERT               ' Find the bit pattern to send to PORTB
       POKE PORTB, Pattern         ' Send Pattern to PORTB
       PAUSE 3000                  ' Wait 3 seconds
       GOTO LOOP

CONVERT:
'
' Find the bit pattern to be sent to PORTB in order to turn on the correct segments
' to display the required number. Dice contains a number between 1 and 6 and
' on return from LOOKUP statement, Pattern contains the bit pattern to send to
' PORTB to display the required number in Dice. Because we are using a
' common-anode display, a segment is turned on when it is logic 0 and thus the
' bit pattern is inverted before sending to PORTB.
'
       LOOKUP Dice, ($3F, $06, $5B, $4F, $66, $6D, $7D, $07, $7F, $6F), Pattern
       Pattern = Pattern ^ $FF          ' Invert bits of variable Pattern
       RETURN


       END                         ' End of program
```

Figure 5.54   PicBasic program of Project 10

PicBasic Pro

The software for PicBasic Pro language is shown in Figure 5.55. The program is very similar to the PicBasic program given in Figure 5.54 with the exception that the registers are addressed directly.

```
'*********************************************************************
'
'              7-SEGMENT DICE
'              ================
'
' In this project a common-anode type 7-segment display is connected
' to PORTB of a PIC16F627 model microcontroller. Additionally, a
' push-button switch is connected to bit 7 of PORTB. When the button
' is pressed, the project displays a number between 1 and 6 just like a
' dice. The RANDOM statement is used to generate a random number.
```

Figure 5.55   (Continued)

```
'
' The microcontroller is operated with the internal 4MHz clock and also
' the internal reset is used.
'
' The connection between the microcontroller and the display is as
' follows:
'
'       RB0      segment a
'       RB1      segment b
'       RB2      segment c
'       RB3      segment d
'       RB4      segment e
'       RB5      segment f
'       RB6      segment g
'
' The decimal point of the display is not used.
'
'
' Author:        Dogan Ibrahim
' Date:          October, 2005
' Compiler:      PicBasic Pro
' File:          LED20.BAS
'
' Modifications
' ===========
'
'*****************************************************************
'
' DEFINITIONS
'
Pattern  VAR   Byte
Dice     VAR   WORD
'
' START OF MAIN PROGRAM
'
         TRISB = $80              ' Bits 0-6 are outputs, bit 7 is input

LOOP:                             ' Display 0 at the beginning
         DICE = 0
         GOSUB CONVERT
         PORTB = Pattern          ' Display 0 to show that we are ready

WT:      RANDOM Dice              ' Generate a random number between 0 and 65535
         IF PORTB.7 = 1 THEN WT   ' Wait until switch is pressed
```

Figure 5.55   (Continued)

```
BR:      Dice = Dice & 7              ' Number between 0 and 7
         IF Dice <> 7 THEN NXT       ' If the number is 0 or 7, get a new number
         RANDOM Dice
         GOTO BR
NXT:
         GOSUB CONVERT               ' Find the bit pattern to send to PORTB
         PORTB = Pattern             ' Send Pattern to PORTB
         PAUSE 3000                  ' Wait 3 seconds
         GOTO LOOP

CONVERT:
'
' Find the bit pattern to be sent to PORTB in order to turn on the correct segments
' to display the required number. Dice contains a number between 1 and 6 and
' on return from LOOKUP statement, Pattern contains the bit pattern to send to
' PORTB to display the required number in Dice. Because we are using a
' common-anode display, a segment is turned on when it is logic 0 and thus the
' bit pattern is inverted before sending to PORTB.
'
         LOOKUP Dice, [$3F, $06, $5B, $4F, $66, $6D, $7D, $07, $7F, $6F], Pattern
         Pattern = Pattern ^ $FF     ' Invert bits of variable Pattern
         RETURN


         END                         ' End of program
```

Figure 5.55   PicBasic Pro program of Project 11

Project 12

Project title: Dual 7-segment LED display

Project description: In this project two 7-segment displays are used. Then, a number (in this case 25) is shown on the displays.

Hardware: When more than one 7-segment display is used the displays are configured and controlled as multiplexed units. Here, as shown in Figure 5.56, the segments of the displays are connected in parallel and their common points are driven separately, each one for a brief period of time. For example, to display number 25, we have to send 2 to the first digit and enable its common point. After a few milliseconds, number 5 is sent to the second digit and the common point of the second digit is enabled. When this process is repeated continuously the user sees as if both displays are on continuously.

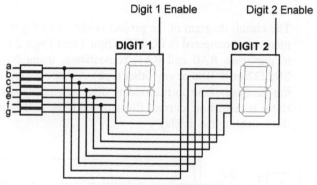

Figure 5.56   Connecting two 7-segment displays in parallel

Some display manufacturers provide multiplexed multi-digit displays in single packages. One such device is the D56 series displays. These are dual red or green colour common-anode or common-cathode displays where the segments of both digits are paralleled and each digit has a separate common control pin. The display used in this project is the D56E05 which is a red colour common-anode two digit display which has the pin configuration as in Table 5.5. This display can be controlled as follows:

- Send the segment data for digit 1 to segments a to g
- Enable digit 1 by connecting digit 1 enable pin to +V supply
- Wait for a few milliseconds
- Send the segment data for digit 2 to segments a to g
- Enable digit 2 by connecting digit 2 enable pin to +V supply
- Wait for a few milliseconds
- Repeat this process continuously

Table 5.5   Pin configuration of D56E05 dual display

Pin number	Segment
1	E
2	D
3	C
4	Digit 1 enable
5	G
6	B
7	A
8	F
9	Digit 2 enable
10	Decimal point

The circuit diagram of the project is shown in Figure 5.57. Display segments are connected to PORTB. Digit 1 and Digit 2 inputs are connected to port pins RA0 and RA1, respectively, using NPN transistors (e.g. 2N2222 or BC108 or any other type). A display digit is enabled by making the base of the corresponding digit transistor logic 1. When the transistor is turned on, current flows through the collector–emitter junction, thus enabling the display.

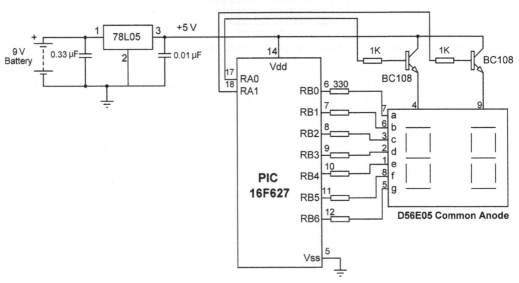

Figure 5.57   Circuit diagram of Project 12

The project constructed on a breadboard is shown in Figure 5.58.

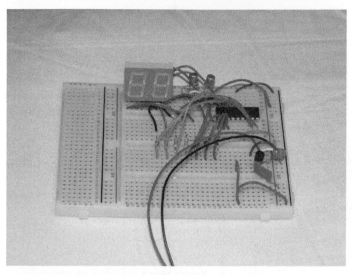

Figure 5.58   Construction of the project

Flow diagram: The flow diagram of the project is shown in Figure 5.59. At the beginning of the program PORTA and PORTB pins are configured as outputs. Variable *Cnt* stores the number to be displayed (loaded with number 25 in this example). First, 10s digit of the display is obtained by dividing *Cnt* by 10. Subroutine CONVERT is then called to obtain the segments to be turned on. This bit pattern is sent to PORTB and then digit 1 is enabled by setting bit 0 of PORTA to logic 1. As a result of this the 10 s digit is displayed. After 1 ms delay the 1 s digit is obtained and the corresponding segment bit pattern is sent to PORTB and then digit 2 is enabled by setting bit 1 of PORTA to logic 1. As a result of this the 1s digit is displayed. The above process is repeated forever.

Software: **PicBasic**
The software for PicBasic language is given in Figure 5.60. At the beginning of the program TRISA and TRISB registers are cleared so that PORTA and PORTB pins are configured as outputs. CMCON register is then set to 7 so that RA0 and RA1 ports are configured as digital I/O. *Cnt*, *Temp*, *Digit*, and *Pattern* are declared as byte variables. Variable *Cnt* is set to number 25 and this is the value we wish to display. *Cnt* is divided by 10 to obtain the 10 s digit of *Cnt* and this number is stored in variable *Digit*.

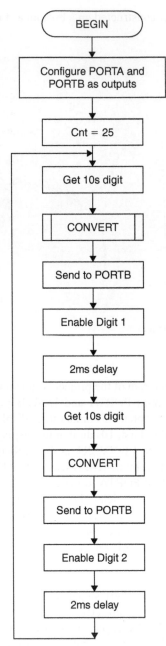

Figure 5.59   Flow diagram of Project 12

```
'****************************************************************
'
'                    DUAL 7-SEGMENT LED DISPLAY
'                    ==============================
'
' In this project two 7-segment LED displays are connected to PORTB
' of a PIC16F627 type microcontroller. The program displays the number
' in variable Cnt on the displays (Cnt is made equal to 25 in this example).
'
' The conenction between the LEDs and the microcontroller are as follows:
'
'          RB0       segment a
'          RB1       segment b
'          RB2       segment c
'          RB3       segment d
'          RB4       segment e
'          RB5       segment f
'          RB6       segment g
'          RA0       digit 1 enable
'          RA1       digit 2 enable
'
' Left digit is Digit 1 and right digit is Digit 2.
'
' The microcontroller operates with a 4MHz internal clock and internal
' power-on reset.
'
'
'
' Author:       Dogan Ibrahim
' Date:         October, 2005
' Compiler:     PicBasic
' File:         LED21.BAS
'
' Modifications
' =============
'
'****************************************************************
'
' SYMBOLS
'
Symbol TRISA = $85                      ' TRISA address
Symbol TRISB = $86                      ' TRISB address
Symbol PORTA = $05                      ' PORTA address
```

Figure 5.60 (Continued)

```
Symbol PORTB = $06                      ' PORTB address
Symbol CMCON = $1F                      ' CMCON address
Symbol Cnt = B0                         ' Cnt is a byte variable
Symbol Temp = B1                        ' Temp is a byte variable
Symbol Digit = B2                       ' Digit is a byte variable
Symbol Pattern = B3                     ' Pattern is a byte variable
'
' START OF MAIN PROGRAM
'
        POKE CMCON, 7                   ' RA0-RA3 are digital I/O
        POKE TRISA, 0                   ' Set PORTA as output
        POKE TRISB, 0                   ' Set all PORTB pins as outputs
        Cnt = 25                        ' Number to display in Cnt
NXT:
        Digit = Cnt / 10                ' Get 10s digit
        GOSUB CONVERT                   ' Get segments to turn on
        POKE PORTB, Pattern             ' Display 10s digit
        POKE PORTA, 1                   ' Enable Digit 1
        PAUSE 2                         ' Wait 2ms

        Temp = Digit * 10
        Digit = Cnt - Temp              ' Get 1s digit
        GOSUB CONVERT                   ' Get segments to turn on
        POKE PORTB, Pattern             ' Display 1s digit
        POKE PORTA, 2                   ' Enable Digit 2
        PAUSE 2                         ' Wait 2ms

        GOTO NXT                        ' Continue displaying

CONVERT:
'
' Find the bit pattern to be sent to PORTB in order to turn on the correct segments
' to display the required number. Digit contains a number between 0 and 9 and
' on return from LOOKUP statement, Pattern contains the bit pattern to send to
' PORTB to display the required number in Digit. Because we are using a
' common-anode display, a segment is turned on when it is logic 0 and thus the
' bit pattern is inverted before sending to PORTB.
'
        LOOKUP Digit, ($3F, $06, $5B, $4F, $66, $6D, $7D, $07, $7F, $6F), Pattern
        Pattern = Pattern ^ $FF         ' Invert bits of variable Pattern
        RETURN

        END                             ' End of program
```

Figure 5.60   PicBasic program of Project 12

Subroutine CONVERT is called to obtain the segments to be turned on. CONVERT receives *Digit* as the input variable and returns variable *Pattern* as the output. *Pattern* is then sent to PORTB and bit 0 of PORTA is set to logic 1 to enable digit 1. After a delay of 1 ms, digit 2 is obtained in variable *Digit*. Subroutine CONVERT is called again to obtain the segments to be turned on. The program then sends the segment pattern to PORTB and sets bit 1 of PORTA to enable digit 2. Program then jumps to label NXT and the process is repeated forever.

PicBasic Pro

The software for PicBasic Pro language is shown in Figure 5.61. At the beginning of the program TRISA and TRISB registers are cleared so that PORTA and PORTB pins are configured as outputs. CMCON register is then set to 7 so that RA0 and RA1 ports are configured as digital I/O. *Cnt*, *Temp*, *Digit*, and *Pattern* are declared as byte variables. Variable *Cnt* is set to number 25 and this is the value we wish to display. PicBasic Pro statement "DIG 1" is used to obtain the first digit of Cnt and subroutine CONVERT is called to obtain the segments to be turned on. CONVERT receives *Digit* as the input variable and returns variable *Pattern* as the output. *Pattern* is then sent to PORTB and bit 0 of PORTA is set to logic 1 to enable digit 1. After a delay of 1 ms, digit 2 is obtained by using the PicBasic Pro statement "DIG 0" and subroutine CONVERT is called again to obtain the segments to be turned on for digit 2. The program then sends the segment pattern to PORTB and sets bit 1 of PORTA to enable digit 2. Program then jumps to label NXT and the process is repeated forever.

```
'*******************************************************************
'
'                    DUAL 7-SEGMENT LED DISPLAY
'                    ==============================
'
' In this project two 7-segment LED displays are connected to PORTB
' of a PIC16F627 type microcontroller. The program displays the number
' in variable Cnt on the displays (Cnt is made equal to 25 in this example).
'
' The conenction between the LEDs and the microcontroller are as follows:
'
'        RB0     segment a
'        RB1     segment b
'        RB2     segment c
```

Figure 5.61   (Continued)

```
'            RB3       segment d
'            RB4       segment e
'            RB5       segment f
'            RB6       segment g
'            RA0       digit 1 enable
'            RA1       digit 2 enable
'
' Left digit is Digit 1 and right digit is Digit 2.
'
' The microcontroller operates with a 4MHz internal clock and internal
' power-on reset.
'
'
'
' Author:          Dogan Ibrahim
' Date:            October, 2005
' Compiler:        PicBasic Pro
' File:            LED22.BAS
'
' Modifications
' ============
'
'*********************************************************************
'
' DEFINITIONS
'
Cnt       VAR       Byte           ' Cnt is a byte variable
Digit     VAR       Byte           ' Digit is a byte variable
Pattern   VAR       Byte           ' Pattern is a byte variable
Digit1    VAR       PORTA.0        ' Digit 1 enable bit
Digit2    VAR       PORTA.1        ' Digit 2 enable bit
'
' START OF MAIN PROGRAM
'
          CMCON = 7                ' RA0-RA3 are digital I/O
          TRISA = 0                ' Set PORTA as output
          TRISB = 0                ' Set all PORTB pins as outputs

          Cnt = 25                 ' Number to display in Cnt
NXT:
          Digit1 = 0               ' Disable digit 1
          Digit2 = 0               ' Disable digit 2
```

Figure 5.61   (Continued)

```
        Digit = Cnt DIG 1              ' Get 10s digit
        GOSUB CONVERT                  ' Get segments to turn on
        PORTB = Pattern                ' Display 10s digit
        Digit1 = 1                     ' Enable Digit 1
        PAUSE 2                        ' Wait 2ms

        Digit = Cnt DIG 0              ' Get 1s digit
        GOSUB CONVERT                  ' Get segments to turn on
        PORTB = Pattern                ' Display 1s digit
        Digit2 = 1                     ' Enable Digit 2
        PAUSE 2                        ' Wait 2ms

        GOTO NXT                       ' Continue displaying


CONVERT:
'
' Find the bit pattern to be sent to PORTB in order to turn on the correct segments
' to display the required number. Digit contains a number between 0 and 9 and
' on return from LOOKUP statement, Pattern contains the bit pattern to send to
' PORTB to display the required number in Digit. Because we are using a
' common-anode display, a segment is turned on when it is logic 0 and thus the
' bit pattern is inverted before sending to PORTB.
'
        LOOKUP Digit, [$3F, $06, $5B, $4F, $66, $6D, $7D, $07, $7F, $6F], Pattern
        Pattern = Pattern ^ $FF        ' Invert bits of variable Pattern
        RETURN


        END                            ' End of program
```

Figure 5.61 PicBasic Pro program of Project 12.

Project 13

Project title: Dual 7-segment LED display counter

Project description: In this project two 7-segment displays are used as in Project 12. The project works like a counter where numbers 00 to 99 are shown on the display with a few seconds delay between each output. The count is repeated after 99.

Hardware: The circuit diagram of the project is as in Project 12 (Figure 5.57). That is, display segments are connected to PORTB, and display digits are controlled from bit 0 and bit 1 of PORT A.

One of the problems in this project is that the display digits require to be updated continuously so that we can see the numbers displayed on each digit. But at the same time we have to increment the count and wait a few seconds before sending a new value to the display. This requires a multi-tasking approach where the display can be updated independent of the counting function. One solution to this problem is to update the display inside a timer interrupt routine which can be done independent of other functions of the program.

The timer interrupt TMR0 can be configured to interrupt at required intervals. When the timer interrupt is enabled and a 4-MHz clock is used, TMR0 interrupt occurs at the time given by T, where T is in microseconds and

$$T = \text{Pre-scaler value} \times (256 - \text{TMR0 value})$$

In this project we shall set the TMR0 to generate interrupts at every 10 ms and this will be our display update time. If we choose a pre-scaler value of 256, the value to be loaded into the TMR0 register is found to be

$$\text{TMR0} = 256 - 10,000/256$$

which is about 217.

Flow diagram: The flow diagram of the project is shown in Figure 5.62. At the beginning of the program PORTA and PORTB pins are configured as outputs, and timer interrupt TMR0 is enabled. The program consists of two sections: the *Main Program* and the *Interrupt Service Routine (ISR)*.

Inside the main program variable *Cnt* is initialised to 0 and the program increments *Cnt* by 1 after every second. When *Cnt* reaches 99, it is cleared again to 0. Subroutine CONVERT is then called to find the segments to be turned on to display a required number.

The display is updated inside the ISR every time a timer interrupt occurs, independent of the main program. Timer register TMR0 is re-loaded with 217 as soon as an interrupt is generated.

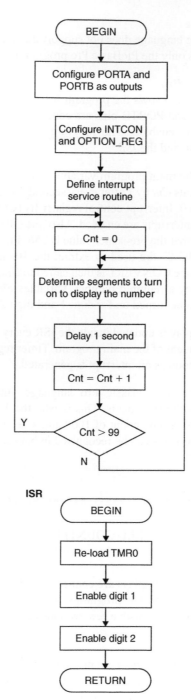

Figure 5.62   Flow diagram of Project 13

Software: **PicBasic**

PicBasic language does not support the use of interrupts from high-level and thus only the PicBasic Pro program is given for this project.

PicBasic Pro

Figure 5.63 shows the program listing. At the beginning of the program PORTA and PORTB pins are configured as outputs, and timer interrupt TMR0 is enabled. The program consists of two sections: the *Main Program*, and the *ISR*.

Inside the main program variable *Cnt* is initialised to 0 and the program increments *Cnt* by 1 after every second. When *Cnt* reaches 99, it is cleared again to 0. Interrupt control register INTCON is configured so that timer TMR0 interrupts are enabled. Also, the OPTION_REG register is configured so that the pre-scaler value is 256. PicBasic Pro statements "DIG 1" and "DIG 0" are used to extract the 10s and the 1s digit of a number. Subroutine CONVERT is then called to find the segments to be turned on to display a required number. Variables *First* and *Second* store the segments to be turned on for digit 1 and digit 2, respectively.

The display is updated inside the ISR every time a timer interrupt occurs, independent of the main program. Timer register TMR0 is re-loaded with 217 as soon as an interrupt is generated.

Note that in PicBasic Pro language, interrupts are only recognised between the statements. This is why the 1 s delay is made up of a FOR loop with a loop count of 1000 and a delay of 1 ms inside the loop. This way, interrupts can be recognised in between the 1 ms delays.

```
'*********************************************************************
'
'               DUAL 7-SEGMENT LED COUNTER
'               ==============================
'
' In this project two 7-segment LED displays are connected to PORTB
' of a PIC16F627 type microcontroller. The program works as a counter
' where numbers 00 to 99 are displayed with a few seconds delay between
' each output.
'
' The program consists of two sections: the main program and the interrupt
' service routine (ISR). The counter increments inside the main program and
' the TMR0 interrupt routine is used to update the displays.
```

Figure 5.63   (Continued)

```
'
' The conenction between the LEDs and the microcontroller are as follows:
'
'         RB0      segment a
'         RB1      segment b
'         RB2      segment c
'         RB3      segment d
'         RB4      segment e
'         RB5      segment f
'         RB6      segment g
'         RA0      digit 1 enable
'         RA1      digit 2 enable
'
' Left digit is Digit 1 and right digit is Digit 2.
'
' The microcontroller operates with a 4MHz internal clock and internal
' power-on reset.
'
'
'
' Author:          Dogan Ibrahim
' Date:            October, 2005
' Compiler:        PicBasic Pro
' File:            LED23.BAS
'
' Modifications
' ===========
'
'*********************************************************************
'
' DEFINITIONS
'
Cnt      VAR    Byte              ' Cnt is a byte variable
Digit    VAR    Byte              ' Digit is a byte variable
Pattern  VAR    Byte              ' Pattern is a byte variable
Digit1   VAR    PORTA.0           ' Digit 1 enable bit
Digit2   VAR    PORTA.1           ' Digit 2 enable bit
First    VAR    Byte              ' First is a byte variable
Second   VAR    Byte              ' Second is a byte variable
i        VAR    Word              ' i is a word variable
```

Figure 5.63   (Continued)

```
'
' START OF MAIN PROGRAM
'
        CMCON = 7                       ' RA0-RA3 are digital I/O
        TRISA = 0                       ' Set PORTA as output
        TRISB = 0                       ' Set all PORTB pins as outputs
'
' Enable TMR0 timer interrupts
'
      INTCON = %00100000              ' Enable TMR0 interrupts
      OPTION_REG = %00000111          ' Initialise the prescale
      TMR0 = 217                      ' Load TMR0 register
      ON INTERRUPT GOTO ISR
      INTCON = %10100000              ' Enable Interrupts

LOOP:
      Cnt = 0                         ' Initialise Cnt to 0
NXT:
      Digit = Cnt DIG 1               ' Get 10s digit
      GOSUB CONVERT                   ' Get segments to turn on
      First = Pattern                 ' Display 10s digit

      Digit = Cnt DIG 0               ' Get 1s digit
      GOSUB CONVERT                   ' Get segments to turn on
      Second = Pattern                ' Display 1s digit

      FOR i = 1 to 1000
          Pause 1                     ' Wait 1 second
      NEXT i

      Cnt = Cnt + 1                   ' Increment Cnt
      IF Cnt > 99 THEN LOOP           ' If Cnt > 99 then goto LOOP
      GOTO NXT                        ' Continue
'
' This is the Interrupt Service Routine (ISR). The program jumps to this
' routine whenever a timer interrupt is generated.
'
DISABLE                               ' Disable further interrupts
ISR:
      TMR0 = 216
      PORTB = First
      Digit2 = 0
      Digit1 = 1
      PAUSE 5
```

Figure 5.63   (Continued)

```
        Digit1 = 0
        PORTB = Second
        Digit2 = 1
        PAUSE 1

        INTCON.2 = 0              ' Re-enable TMR0 interrupts
        RESUME                   ' Return to main program
        ENABLE                   ' Enable interrupts
CONVERT:
'
' Find the bit pattern to be sent to PORTB in order to turn on the correct segments
' to display the required number. Digit contains a number between 0 and 9 and
' on return from LOOKUP statement, Pattern contains the bit pattern to send to
' PORTB to display the required number in Digit. Because we are using a
' common-anode display, a segment is turned on when it is logic 0 and thus the
' bit pattern is inverted before sending to PORTB.
'
        LOOKUP Digit, [$3F, $06, $5B, $4F, $66, $6D, $7D, $07, $7F, $6F], Pattern
        Pattern = Pattern ^ $FF          ' Invert bits of variable Pattern
        RETURN

        END                              ' End of program
```

Figure 5.63   PicBasic Pro program of Project 13

Project 14

Project title: Dual 7-segment LED event counter

Project description: In this project two 7-segment displays are used as in Project 13. A push-button switch is connected to bit 7 of PORTB. The project counts and displays the number of times the switch is pressed. This project can be used to count events in many other applications, such as counting the number of products passing on a conveyor belt, number of people entering a building, number of cars entering a car park, and so on.

Hardware: The circuit diagram of the project is similar to Figure 5.57, but here, as shown in Figure 5.64, a push-button switch is connected to bit 7 of PORTB. A pull-up resistor is used so that the switch is normally at logic 1 and goes to logic 0 when it is pressed.

The timer TMR0 interrupt as in Project 13 is used to display the count on the dual 7-segment display.

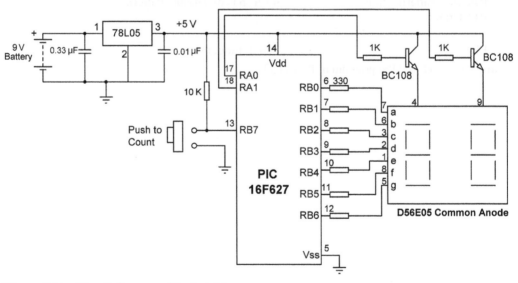

Figure 5.64 Circuit diagram of Project 14

Flow diagram: The flow diagram of the project is shown in Figure 5.65. At the beginning of the program PORTA and PORTB pins are configured as outputs, and timer interrupt TMR0 is enabled. The program consists of two sections: the *Main Program*, and the *ISR*.

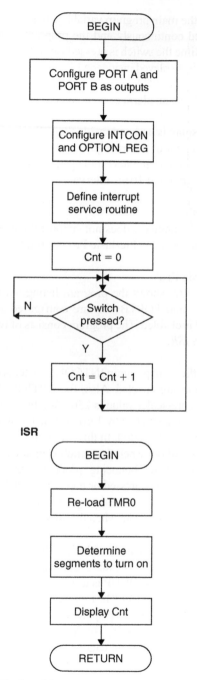

Figure 5.65   Flow diagram of Project 14

Inside the main program variable *Cnt* is initially cleared to 0 and the switch is tested continuously and the program waits until the switch is pressed. Every time the switch is pressed variable *Cnt* is incremented by 1. The value of *Cnt* is displayed inside the ISR every 10 ms. In a practical application the program should check to make sure that *Cnt* does not become greater than 99 and give an alarm or some other warning before this happens.

The display is updated inside the ISR every time a timer interrupt occurs, independent of the main program. Timer register TMR0 is re-loaded with 217 as soon as an interrupt is generated. As in Project 13, timer interrupts are generated at 10 ms intervals.

Software:

PicBasic
PicBasic language does not support the use of interrupts from high-level and thus only the PicBasic Pro program is given for this project.

PicBasic Pro
Figure 5.66 shows the program listing. At the beginning of the program PORTA and PORTB pins are configured as outputs, and timer interrupt TMR0 is enabled. The program consists of two sections: the *Main Program* and the *ISR*.

Inside the main program INTCON register is configured so that timer TMR0 interrupts are enabled. Also, the OPTION_REG register is configured so that the pre-scaler value is 256. Variable *Cnt* is initialised to 0 and the program increments *Cnt* by 1 every time the switch is pressed (or whenever an external event occurs). In this program PicBasic Pro statement BUTTON is used to find out when the switch is pressed. The statement is configured to eliminate switch-bouncing problems. Switch-contact bouncing happens when a switch is pressed or released. Switch contacts oscillate and generate noise which may cause the microcontroller to read multiple on/off readings or wrong switch state when the switch is pressed or released.

The display is updated inside the ISR. PicBasic Pro statements "DIG 1" and "DIG 0" are used to extract the 10 s and the 1s digits of variable *Cnt*. PicBasic Pro statement LOOKUP is used to find the segments to be turned on to display each digit of *Cnt*. The bits are inverted before they are sent to the display. This is done because a display segment is turned on when a logic 0 is applied (common-anode display) to the segment. The inversion is done by bit-wise exclusive-or'ing the bit data with hexadecimal number $FF (a bit exclusive-or'ed with 1 is inverted). At the beginning of the ISR, timer register TMR0 is re-loaded with 217 so that the next interrupt is generated after 10 ms.

```
'***************************************************************************
'
'                    DUAL 7-SEGMENT EVENT COUNTER
'                    ================================
'
' In this project two 7-segment LED displays are connected to PORTB
' of a PIC16F627 type microcontroller. Also, a push-button switch is
' connected to bit 7 of PORTB. The program counts and displays how
' many times the switch is pressed. Although a simple switch is used
' in this example, the project can be used to count events such as the
' number of objects passing on a conveyor belt, number of cars entering
' a car park etc.
'
' In this project the switch is debounced to eliminate the contact problems
' using the PicBasic Pro BUTTON statement.
'
' The program consists of two sections: the main program and the interrupt
' service routine (ISR). The counter increments inside the main program and
' the TMR0 interrupt routine is used to update the displays.
'
' The conenction between the LEDs and the microcontroller are as follows:
'
'        RB0      segment a
'        RB1      segment b
'        RB2      segment c
'        RB3      segment d
'        RB4      segment e
'        RB5      segment f
'        RB6      segment g
'        RA0      digit 1 enable
'        RA1      digit 2 enable
'
'        RB7      push-button switch
'
' Left digit is Digit 1 and right digit is Digit 2.
'
' The microcontroller operates with a 4MHz internal clock and internal
' power-on reset.
'
'
'
' Author:      Dogan Ibrahim
' Date:        October, 2005
' Compiler:    PicBasic Pro
' File:        LED24.BAS
```

Figure 5.66 (Continued)

```
'
' Modifications
' ============
'
'
'**********************************************************************
'
' DEFINITIONS
'
Cnt      VAR    Byte                    ' Cnt is a byte variable
Digit    VAR    Byte                    ' Digit is a byte variable
Pattern  VAR    Byte                    ' Pattern is a byte variable
Digit1   VAR    PORTA.0                 ' Digit 1 enable bit
Digit2   VAR    PORTA.1                 ' Digit 2 enable bit
Pbutton  VAR    PORTB.7                 ' Push button is bit 7 of PORTB
i        VAR    Byte                    ' i is a byte variable
'
' START OF MAIN PROGRAM
'
      CMCON = 7                         ' RA0-RA3 are digital I/O
      TRISA = 0                         ' Set PORTA as output
      TRISB = $80                       ' Bit 7 of PORTB input, others outputs
'
' Enable TMR0 timer interrupts
'
      INTCON = %00100000                ' Enable TMR0 interrupts
      OPTION_REG = %00000111            ' Initialise the prescale
      TMR0 = 217                        ' Load TMR0 register
      ON INTERRUPT GOTO ISR
      INTCON = %10100000                ' Enable Interrupts

      Cnt = 0                           ' Initialise event counter to 0

LOOP:
      BUTTON Pbutton, 0, 255,0, i, 0, LOOP   ' Wait until push-button is pressed and
debounce switch
      Cnt = Cnt + 1                     ' Increment event counter
      GOTO LOOP                         ' Continue
'
' This is the Interrupt Service Routine (ISR). The program jumps to this
' routine whenever a timer interrupt is generated. Inside this routine the
' value of variable Cnt is displayed.
'
DISABLE                                 ' Disable further interrupts
ISR:
      TMR0 = 217
```

Figure 5.66   (Continued)

```
Digit = Cnt DIG 1                                    ' Get 10s digit
LOOKUP Digit, [$3F, $06, $5B, $4F, $66, $6D, $7D, $07, $7F, $6F], Pattern
Pattern = Pattern ^ $FF                              ' Invert bits of variable Pattern
PORTB = Pattern                                      ' Display 10s digit
Digit2 = 0                                           ' Disable digit 2
Digit1 = 1                                           ' Enable digit 1
Pause 5                                              ' Wait 5ms

Digit = Cnt DIG 0                                    ' Get 1s digit
LOOKUP Digit, [$3F, $06, $5B, $4F, $66, $6D, $7D, $07, $7F, $6F], Pattern
Pattern = Pattern ^ $FF                              ' Invert bits of variable Pattern
Digit1 = 0                                           ' Disable digit 1
PORTB = Pattern                                      ' Display 1s digit
Digit2 = 1                                           ' Enable digit 2
PAUSE 1                                              ' Wait 1ms
INTCON.2 = 0                                         ' Re-enable TMR0 interrupts
RESUME                                               ' Return to main program
ENABLE                                               ' Enable interrupts

END                                                  ' End of program
```

Figure 5.66   PicBasic Pro program of Project 14

Project 15

Project title: 4-digit LED display with serial driver – counter project

Project description: In this project a 4-digit serial 7-segment display is used as a decimal counter. The display counts up by one every second from 0000 to 9999. When it reaches 9999, it goes back to 0000 and the process continues forever.

Hardware: Multiplexed 7-segment displays are so important in many display-based applications that several manufacturers have designed multi-digit, multiplexed displays with built-in drivers. One such display is the 4-digit multiplexed 7-segment display B08M04N, manufactured by *Nexus Machines Ltd*. This is a family of displays with sizes ranging from 8 to 38 mm and available in red, green, and yellow colours.

In this project, a B08M04N-R red colour 8 mm 4-digit 7-segment display is used. Figure 5.67 shows the picture of this display.

Figure 5.67   B08M04N-R display

The display has 9 pins as shown in Table 5.6. The on-board driver chip has a serial input format that features serial data, clock and chip enable. A single +5 V supply is normally used, although the unit will work with a supply as high as +10 V. Serial data is sent as 36 bits of segment information where a logic 1 turns a segment ON. The displays have 2 spare outputs that can be used for driving external LEDs, where the LED current is programmed via an on-board resistor.

Table 5.6 B08M04N display pin configuration

Pin number	Description
1	LED 1 drive
2	LED 2 drive
3	Chip enable
4	Data
5	Clock
6	Vdd (+5 V)
7	Brightness
8	Gnd (0 V)
9	Vled

Table 5.7 shows how data should be sent to the display unit. First, a start bit (logic 1) is sent. After this, the segments a to g and the decimal point of each digit are sent consecutively starting from digit 1, which is the digit at the right-most position. The start bit and the 4-digit display data is sent in 33 bits. Then the bits for the 2 LED are sent (a logic 1 turns on an LED). The last bit sent is a NULL bit.

Table 5.7 Display segment data

Bit 0	START	Bit 9	A2	Bit 17	A3	Bit 25	A4	Bit 33	LED 1
Bit 1	A1	Bit 10	B2	Bit 18	B3	Bit 26	B4	Bit 34	LED2 2
Bit 2	B1	Bit 11	C2	Bit 19	C3	Bit 27	C4	Bit 35	Null
Bit 3	C1	Bit 12	D2	Bit 20	D3	Bit 28	D4		
Bit 4	D1	Bit 13	E2	Bit 21	E3	Bit 29	E4		
Bit 5	E1	Bit 14	F2	Bit 22	F3	Bit 30	F4		
Bit 6	F1	Bit 15	G2	Bit 23	G3	Bit 31	G4		
Bit 7	G1	Bit 16	DP2	Bit 24	DP3	Bit 32	DP4		
Bit 8	DP1								

The display control is summarised below (note that each bit should be followed by a clock bit):

- Send START bit (logic 1)
- Send A1 to G1 of digit 1 (right-most digit)
- Send decimal point (DP1) of digit 1

- Send A2 to G2 of digit 2
- Send decimal point (DP2) of digit 2
- Send A3 to G3 of digit 3
- Send decimal point (DP3) of digit 3
- Send A4 to G4 of digit 4 (left-most digit)
- Send decimal point (DP4) of digit 4
- Send LED1 bit
- Send LED2 bit
- Send a NULL bit.

The relationship between a digit number and the segments to be turned on to display this number is given in Table 5.8. For example, to display number 4 in a digit, we have to send the hexadecimal number $66 to the digit, i.e. bit pattern "01100110". The segment of each digit must be sent first, i.e. the bits must be shifted left as they are sent to the display. A display is blank if a 0 is sent to all of its segments. This can be useful when we want to turn off a leading zero when displaying a number. For example, number "23" can be displayed as "0023" or "023" or as "23" where the spaces correspond to blank characters. Leading zeroes are usually not shown in displays and the correct format is "23".

Table 5.8   Relationship between segments and numbers

Number to display	a b c d e f g dp	Number (Hex)
0	1 1 1 1 1 1 0 0	$FC
1	0 1 1 0 1 1 1 1	$60
2	1 1 0 1 1 0 1 0	$DA
3	1 1 1 1 0 0 1 0	$F2
4	0 1 1 0 0 1 1 0	$66
5	1 0 1 1 0 1 1 0	$B6
6	1 0 1 1 1 1 1 0	$BE
7	1 1 1 0 0 0 0 0	$E0
8	1 1 1 1 1 1 1 0	$FE
9	1 1 1 1 0 1 1 0	$F6

As an example, suppose that we wish to display number 2478 with both LEDs turned off. The following data should then be sent to the displays:

- Send a START bit
- Send bit pattern of hexadecimal $FE (i.e. "11111110") with the MSB bit sent first to display 8 on digit 1

- Send bit pattern of hexadecimal $E0 (i.e. "11100000") with the MSB bit sent first to display 7 on digit 2
- Send bit pattern of hexadecimal $66 (i.e. "01100110") with the MSB bit sent first to display 4 on digit 3
- Send bit pattern of hexadecimal $DA (i.e. "11011010") with the MSB bit sent first to display 2 on digit 4
- Send 0 for LED 1 to turn off LED 1
- Send 0 for LED 2 to turn off LED 2
- Send 0 as the NULL bit.

That is, the following 36 bits should be sent to the display with a clock bit sent after each bit (a space character is used between each digit data for clarity):

"1 11111110 11100000 01100110 11011010 0 0 0"[s1]

Similarly, for example, number 34 with leading zeroes and with both LEDs turned on can be displayed by sending the following bit pattern to the display:

- Send a START bit
- Send bit pattern of hexadecimal $66 (i.e. "01100110") with the MSB bit sent first to display 4 on digit 1
- Send bit pattern of hexadecimal $F2 (i.e. "11110010") with the MSB bit sent first to display 3 on digit 2
- Send bit pattern 0 (i.e. "00000000") to blank digit 3
- Send bit pattern 0 (i.e. "00000000") to blank digit 4
- Send 1 for LED 1 to turn on LED 1
- Send 1 for LED 2 to turn on LED 2
- Send 0 as the NULL bit terminator.

That is, the following 36 bits should be sent to the display with a clock bit sent after each bit (a space character is used between each digit data for clarity):

"1 01100110 11110010 00000000 00000000 1 1 0"[s2]

The circuit diagram of Project 15 is shown in Figure 5.68. In this project a PIC16F627-type microcontroller is used with 4 MHz internal clock and internal power-on reset. Display data and clock are connected to bit 6 (RB6) and bit 7 (RB7) of PORTB, respectively.

The project built on a breadboard is shown in Figure 5.69. Note that the project is very simple and consists of only a few connections.

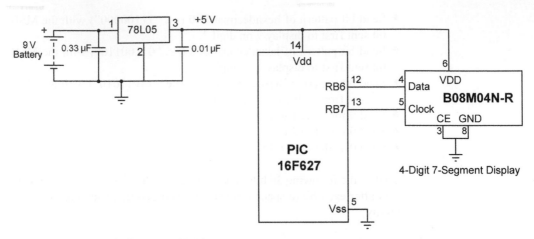

Figure 5.68   Circuit diagram of Project 15

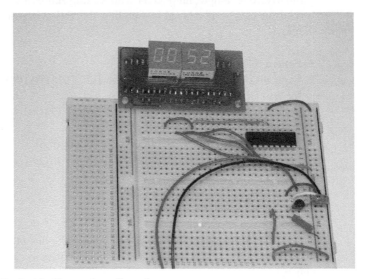

Figure 5.69   Project built on a breadboard

Flow diagram: The flow diagram of the project is shown in Figure 5.70. At the beginning of the program PORTB pins are configured as outputs and the states of LED 1 and LED 2 are set as required. Then the bit pattern to be sent to the display to show the value of variable *Cnt* is determined and this data is sent to the display. The program then waits for 1 s, increments *Cnt* by one, and this process is repeated forever. Thus, the display shows 000 001 002 … 998 999 000 001 … .

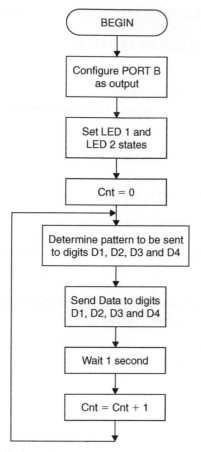

Figure 5.70   Flow diagram of Project 15

Software:

PicBasic

Figure 5.71 shows the PicBasic program listing. At the beginning of the program various program variables are configured as bytes or words. The main program starts by clearing TRISB register so that PORTB pins are configured as outputs. Variable *Cnt* is also cleared to zero since the count will start from 0. The program loop starts with label NXT. Here, the 4 digits of variable *Cnt* are extracted by dividing *Cnt* repeatedly by the powers of 10 and taking the decimal value and the remainder (PicBasic does not support the DIG statement which is available only on PicBasic Pro). After the digits are obtained, subroutine CONVERT is called to find the 7-segment bit pattern of each digit. Variables D1, D2, D3, and D4 store the bit patterns to be sent to each digit of the display.

```
'******************************************************************
'
'                 4-DIGIT 7-SEGMENT LED DISPLAY
'                 =============================
'
' In this project a B08M04 type 4-digit 7-segment LED displays is used.
' The program counts up by one every second. LED 1 and LED 2 are turned
' off in this example.
'
' The display digits are organised as follows:
'
'         D4        D3        D2        D1
'
' Data is sent: D1 first, then D2, then D3 and finally D4
'
' A PIC16F627 type microcontroller is used in the project with 4MHz
' internal clock and internal reset.
'
' The connection between the display and the microcontroller is as follows:
' (display CE pin is connected to ground permanently)
'
'         RB6       Display DATA
'         RB7       Display CLOCK
'
'
'
' Author:          Dogan Ibrahim
' Date:            October, 2005
' Compiler:        PicBasic
' File:            LED25.BAS
'
' Modifications
'=============
'
'******************************************************************
'
' SYMBOLS
'
Symbol TRISB = $86                        ' TRISB address
Symbol PORTB = $06                        ' PORTB address
Symbol Pattern = B0                       ' Pattern is a byte variable
Symbol I = B1                             ' Loop counter variable
Symbol Digit = B2                         ' Digit is a byte variable
```

Figure 5.71   (Continued)

```
Symbol D1 = B3                         ' Digit 1 pattern
Symbol D2 = B4                         ' Digit 2 pattern
Symbol D3 = B5                         ' Digit 3 pattern
Symbol D4 = B6                         ' Digit 4 pattern
Symbol LED1 = B7                       ' Display LED 1 data
Symbol LED2 = B8                       ' Display LED 2 data
Symbol Cnt = W6                        ' Cnt is a word variable
Symbol Temp = W7                       ' Temp is a word variable
Symbol DATA = Pin6                     ' Display Data is RB6
Symbol CLK = 7                         ' Display CLOCK is RB7
'
' START OF MAIN PROGRAM
'
        POKE TRISB, 0                  ' Set PORTB as output

        LED1 = 0                       ' LED 1 is to be off
        LED2 = 0                       ' LED 2 is to be off

        Cnt = 0                        ' Number to display in Cnt
NXT:
        Digit = Cnt / 1000             ' Get 1000s digit
        GOSUB CONVERT                  ' Get segments to turn on
        D4 = Pattern                   ' Pattern for 1000s digit

        Temp = Cnt // 1000             ' Find remainder
        Digit = Temp / 100             ' Get 100s digit
        GOSUB CONVERT                  ' Get segments to turn on
        D3 = Pattern                   ' Pattern for 100s digit

        Temp = Cnt // 100              ' Find remainder
        Digit = Temp / 10              ' Get 10s digit
        GOSUB CONVERT                  ' Get segments to turn on
        D2 = Pattern                   ' Pattern for 10s digit

        Digit = Temp // 10             ' Find remainder
        GOSUB CONVERT                  ' get segments to turn on
        D1 = Pattern                   ' Pattern for 1s digit
'
' Send data to the display
'
        GOSUB SEND_START               ' Send START bit
        Pattern = D1
```

Figure 5.71   (Continued)

```
        GOSUB DISPLAY                      ' Send 1s digit
        Pattern = D2
        GOSUB DISPLAY                      ' Send 10s digit
        Pattern = D3
        GOSUB DISPLAY                      ' Send 100s digit
        Pattern = D4
        GOSUB DISPLAY                      ' Send 1000s digit
        GOSUB SEND_LEDS                    ' Send LED bits
        GOSUB SEND_TERM                    ' Send TERMINATOR bit

        PAUSE 1000                         ' Wait 1 second

        Cnt = Cnt + 1                      ' Increment count

        GOTO NXT                           ' Continue counting and displaying

CONVERT:
'
' Find the bit pattern to be sent to the display in order to turn on the correct segments
' to display the required number. Digit contains a number between 0 and 9 and
' on return from LOOKUP statement, Pattern contains the bit pattern to send to

' PORTB to display the required number in Digit.
'
        LOOKUP Digit, ($FC, $60, $DA, $F2, $66, $B6, $BE, $E0, $FE, $F6), Pattern
        RETURN

SEND_START:
'
' This subroutine sends a START bit to the display. START bit is a logic 1
'
        DATA = 1                           ' Data = 1
        TOGGLE CLK                         ' CLK = 1
        TOGGLE CLK                         ' CLK = 0
        RETURN

SEND_TERM:
'
' This subroutine sends a TERMINATOR bit to the display. TERMINATOR bit is a logic 0
'
        DATA = 0                           ' Data = 0
        TOGGLE CLK                         ' CLK = 1
        TOGGLE CLK                         ' CLK = 0
        RETURN
```

Figure 5.71   (Continued)

```
SEND_LEDS:
'
' This subroutine sends the two LED data to the display
'
        DATA = LED1
        TOGGLE CLK
        TOGGLE CLK
        DATA = LED2
        TOGGLE CLK
        TOGGLE CLK
        RETURN

DISPLAY:
'
' This subroutine sends data and clock bits to the display. Data bits are sent by left shifting
' the value in variable Pattern. A clock pulse is sent after sending each data bit.
'
        FOR I = 1 TO 8
            DATA = Bit7                         ' Get bit 7 of Pattern
            TOGGLE CLK                          ' CLK = 1
            Pattern = Pattern * 2               ' Shift left pattern 1 digit
            TOGGLE CLK                          ' CLK = 0
        NEXT I
        RETURN

        END                                     ' End of program
```

Figure 5.71   PicBasic program of Project 15

Subroutine SEND_START is called to send the START bit to the display. Then the bit pattern of each digit is sent, starting with digit 1. After sending the four-digit data, subroutine SEND_LEDS is called to send the two LED bit data. Data transfer is complete when the terminator NULL character is sent by calling subroutine SEND_TERM.

The program given in Figure 5.71 can be improved and made easier to understand if the display subroutines are all collected and stored inside a common subroutine. This is shown in Figure 5.72. Here, a subroutine called DISPLAY is created and all the display related programs are stored inside this subroutine. The main program consists of the counter *Cnt* only which is incremented every second. Subroutine DISPLAY is then called to display the value of *Cnt*. The advantage of this approach is that the DISPLAY subroutine can be used in other programs after it has been tested and working correctly.

```
'********************************************************************
'
'                    4-DIGIT 7-SEGMENT LED DISPLAY
'                    ===============================
'
' In this project a B08M04 type 4-digit 7-segment LED displays is used.
' The program counts up by one every second. LED 1 and LED 2 are turned
' off in this example.
'
' The display digits are organised as follows:
'
'         D4        D3        D2        D1
'
' Data is sent: D1 first, then D2, then D3 and finally D4
'
' A PIC16F627 type microcontroller is used in the project with 4MHz
' internal clock and internal reset.
'
' The connection between the display and the microcontroller is as follows:
' (display CE pin is connected to ground permanently)
'
'         RB6       Display DATA
'         RB7       Display CLOCK
'
'
'
' Author:        Dogan Ibrahim
' Date:          October, 2005
' Compiler:      PicBasic
' File:          LED26.BAS
'
' Modifications
'==============
'
'********************************************************************
'
' SYMBOLS
'
Symbol TRISB = $86                    ' TRISB address
Symbol PORTB = $06                    ' PORTB address
Symbol Pattern = B0                   ' Pattern is a byte variable
Symbol I = B1                         ' Loop counter variable
Symbol Digit = B2                     ' Digit is a byte variable
```

Figure 5.72   (Continued)

```
Symbol D1 = B3                          ' Digit 1 pattern
Symbol D2 = B4                          ' Digit 2 pattern
Symbol D3 = B5                          ' Digit 3 pattern
Symbol D4 = B6                          ' Digit 4 pattern
Symbol LED1 = B7                        ' Display LED 1 data
Symbol LED2 = B8                        ' Display LED 2 data
Symbol Cnt = W6                         ' Cnt is a word variable
Symbol Temp = W7                        ' Temp is a word variable
Symbol DATA = Pin6                      ' Display Data is RB6
Symbol CLK = 7                          ' Display CLOCK is RB7
'
' START OF MAIN PROGRAM
'
        POKE TRISB, 0                   ' Set PORTB as output

        LED1 = 0                        ' LED 1 is to be off
        LED2 = 0                        ' LED 2 is to be off
        Cnt = 0                         ' Number to display in Cnt

NXT: GOSUB DISPLAY                       ' Display number in Cnt
        PAUSE 1000                      ' Wait 1 second
        Cnt = Cnt + 1                   ' Increment count
        GOTO NXT                        ' Continue counting and displaying

'============================SUBROUTINES ============================
DISPLAY:
'
' This subroutine displays the number in variable Cnt on the 4-digit 7-segment display
'
        Digit = Cnt / 1000              ' Get 1000s digit
        GOSUB CONVERT                   ' Get segments to turn on
        D4 = Pattern                    ' Pattern for 1000s digit

        Temp = Cnt // 1000              ' Find remainder
        Digit = Temp / 100              ' Get 100s digit
        GOSUB CONVERT                   ' Get segments to turn on
        D3 = Pattern                    ' Pattern for 100s digit

        Temp = Cnt // 100               ' Find remainder
        Digit = Temp / 10               ' Get 10s digit
        GOSUB CONVERT                   ' Get segments to turn on
        D2 = Pattern                    ' Pattern for 10s digit
```

Figure 5.72   (Continued)

```
        Digit = Temp // 10            ' Find remainder
        GOSUB CONVERT                 ' get segments to turn on
        D1 = Pattern                  ' Pattern for 1s digit
'
' Send data to the display
'
        GOSUB SEND_START              ' Send START bit
        Pattern = D1
        GOSUB SEGMENTS                ' Send 1s digit
        Pattern = D2
        GOSUB SEGMENTS                ' Send 10s digit
        Pattern = D3
        GOSUB SEGMENTS                ' Send 100s digit
        Pattern = D4
        GOSUB SEGMENTS                ' Send 1000s digit
        GOSUB SEND_LEDS               ' Send LED bits
        GOSUB SEND_TERM               ' Send TERMINATOR bit
        RETURN

CONVERT:
'
' Find the bit pattern to be sent to the display in order to turn on the correct segments
' to display the required number. Digit contains a number between 0 and 9 and
' on return from LOOKUP statement, Pattern contains the bit pattern to send to
' PORTB to display the required number in Digit.
'
        LOOKUP Digit, ($FC, $60, $DA, $F2, $66, $B6, $BE, $E0, $FE, $F6), Pattern
        RETURN

SEND_START:
'
' This subroutine sends a START bit to the display. START bit is a logic 1
'
        DATA = 1                      ' Data = 1
        TOGGLE CLK                    ' CLK = 1
        TOGGLE CLK                    ' CLK = 0
        RETURN

SEND_TERM:
'
' This subroutine sends a TERMINATOR bit to the display. TERMINATOR bit is a logic 0
'
        DATA = 0                      ' Data = 0
        TOGGLE CLK                    ' CLK = 1
        TOGGLE CLK                    ' CLK = 0
        RETURN
```

Figure 5.72   (Continued)

SEND_LEDS:
'
' This subroutine sends the two LED data to the display
'

```
        DATA = LED1
        TOGGLE CLK
        TOGGLE CLK
        DATA = LED2
        TOGGLE CLK
        TOGGLE CLK
        RETURN
```

SEGMENTS:
'
' This subroutine sends data and clock bits to the display. Data bits are sent by left shifting
' the value in variable Pattern. A clock pulse is sent after sending each data bit.
'

```
        FOR I = 1 TO 8
            DATA = Bit7              ' Get bit 7 of Pattern
            TOGGLE CLK               ' CLK = 1
            Pattern = Pattern * 2    ' Shift left pattern 1 digit
            TOGGLE CLK               ' CLK = 0
        NEXT I
        RETURN

        END                         ' End of program
```

Figure 5.72   Improved PicBasic program of Project 15

PicBasic Pro

Figure 5.73 shows the program listing. The PicBasic Pro program is much smaller and also easier to understand than the PicBasic program. The digits of variable *Cnt* are found using the PicBasic Pro DIG statement. DIG 0 returns the 1s digit of a variable, DIG 1 returns the 10s digit and so on.

PicBasic Pro also supports the SHIFTOUT statement which is used to send data and clock bits to the display. The Mode parameter of SHIFTOUT statement is chosen 1 so that the data is shifted out highest bit first.

The display related code is stored inside a subroutine called DISPLAY. Main program consists of the counter *Cnt* only which is incremented every second and subroutine DISPLAY is called to display its value.

```
'********************************************************************
'
'                    4-DIGIT 7-SEGMENT LED DISPLAY
'                    ==============================
'
' In this project a B08M04 type 4-digit 7-segment LED displays is used.
' The program counts up by one every second. LED 1 and LED 2 are turned
' off in this example.
'
' The display digits are organised as follows:
'
'         D4      D3      D2      D1
'
' Data is sent: D1 first, then D2, then D3 and finally D4
'
' A PIC16F627 type microcontroller is used in the project with 4MHz
' internal clock and internal reset.
'
' The connection between the display and the microcontroller is as follows:
' (display CE pin is connected to ground permanently)
'
'         RB6     Display DATA
'         RB7     Display CLOCK
'
' In this program the leftmost LEDs which are zero are blanked so that for example
' number 25 is displayed as "25" and not as "0025"
'
'
' Author:        Dogan Ibrahim
' Date:          October, 2005
' Compiler:      PicBasic Pro
' File:          LED27.BAS
'
' Modifications
' =============
'
'********************************************************************
'
' DEFINITIONS
'
Pattern   VAR Byte                      ' Pattern is a byte variable
I         VAR Byte                      ' Loop counter variable
Digit     VAR Byte                      ' Digit is a byte
```

Figure 5.73   (Continued)

```
LED1     VAR Bit                        ' Display LED 1 data
LED2     VAR Bit                        ' Display LED 2 data
Cnt      VAR Word                       ' Cnt is a word variable
Symbol DATA_PIN = PORTB.6              ' Display Data is RB6
Symbol CLK_PIN = PORTB.7              ' Display CLOCK is RB7

'

' START OF MAIN PROGRAM
'

        TRISB = 0                      ' Set PORTB as output
        LED1 = 0                       ' LED 1 is to be off
        LED2 = 0                       ' LED 2 is to be off
        Cnt = 0                        ' Number to display in Cnt

NXT: GOSUB DISPLAY                     ' Display number in Cnt
        PAUSE 1000                     ' Wait 1 second
        Cnt = Cnt + 1                  ' Increment count
        GOTO NXT                       ' Continue counting and displaying
'========================= SUBROUTINES =========================
DISPLAY:
'

' This subroutine displays the number in variable Cnt on the 4-digit 7-segment display
'

' Send START bit
'

        DATA_PIN = 1                   ' Data = 1
        PULSOUT CLK_PIN, 1             ' Send a clock
'

' Send digit bits
'

        FOR I = 0 TO 3
           Digit = Cnt DIG I            ' Get digits of variable Cnt
           LOOKUP Digit, [$FC, $60, $DA, $F2, $66, $B6, $BE, $E0, $FE, $F6], Pattern
           SHIFTOUT DATA_PIN, CLK_PIN, 1, [ Pattern ]
        NEXT I

'

' Send LED1 and LED 2 bits
'

        DATA_PIN = LED1                ' Data = LED1
        PULSOUT CLK_PIN,1              ' Send clock
        DATA_PIN = LED2                ' Data = LED 2
        PULSOUT CLK_PIN,1              ' Send clock
```

Figure 5.73   (Continued)

```
'
' Send TERMINATOR bit
'
        DATA_PIN = 0              ' Data = 0
        PULSOUT CLK_PIN,1         ' Send clock
        RETURN

        END                       ' End of program
```

Figure 5.73   PicBasic Pro program of Project 15

Project 16

Project title: 4-digit LED display with serial driver – counter project with leading zeroes blanked

Project description: This project is very similar to Project 15 where a 4-digit 7-segment display is used as a counter. In this project, the leading zeroes of the display are blanked. Thus, for example, number "67" is displayed as "67", number "5" is displayed as "5", and so on.

Hardware: The hardware and the circuit diagram of the project is as in Figure 5.68 where the display is controlled from bit 6 and bit 7 of PORTB.

Flow diagram: The flow diagram of the project is very similar to the flow diagram given in Figure 5.70. Here, the difference is that the leading zeroes are blanked by sending zeroes to all of their segments. Figure 5.74 shows the flow diagram of the project.

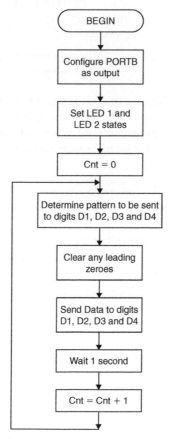

Figure 5.74 Flow diagram of Project 16

Software: **PicBasic**

Figure 5.75 shows the PicBasic program listing. The program is very similar to the one given in Figure 5.72. Here, the values of leading digits are checked and if they are zero, the segments of these digits are cleared to zeroes. Leading zero checking is done by introducing the following code just before sending the segment data to the display:

```
                      ..............................
                      ..............................
                      IF D4 = $FC THEN BL4
                      GOTO CONT
           BL4:    D4 = 0
                      IF D3 = $FC THEN BL3
                      GOTO CONT
           BL3:    D3 = 0
                      IF D2 = $FC THEN BL2
                      GOTO CONT
           BL2:    D2 = 0
           CONT:
                      ....................
                      ....................
```

If digit 4 bit pattern (D4) is equal to hexadecimal $FC then this digit is zero and since it is the left-most digit, it is blanked by clearing D4. If both D4 and D3 bit patterns are zero then both displays are blanked. Finally, if D4, D3, and D2 bit patterns are zero then all three digits are blanked.

```
'*********************************************************************
'
'
'              4-DIGIT 7-SEGMENT LED DISPLAY COUNTER WITH BLANKING
'            ===============================================================
'
'
' In this project a B08M04 type 4-digit 7-segment LED displays is used.
' The program counts up by one every second. LED 1 and LED 2 are turned
' off in this example.
'
' The display digits are organised as follows:
'
'        D4      D3      D2      D1
'
' Data is sent: D1 first, then D2, then D3 and finally D4
```

Figure 5.75   (Continued)

```
' A PIC16F627 type microcontroller is used in the project with 4MHz
' internal clock and internal reset.
'
' The connection between the display and the microcontroller is as follows:
' (display CE pin is connected to ground permanently)
'
'        RB6        Display DATA
'        RB7        Display CLOCK
'
' In this program the leftmost LEDs which are zero are blanked so that for example
' number 25 is displayed as "25" and not as "0025"
'
'
' Author:        Dogan Ibrahim
' Date:          October, 2005
' Compiler:      PicBasic
' File:          LED28.BAS
'
' Modifications
' =============
'
'*********************************************************************
'
' SYMBOLS
'
Symbol TRISB = $86              ' TRISB address
Symbol PORTB = $06              ' PORTB address
Symbol Pattern = B0             ' Pattern is a byte variable
Symbol I = B1                   ' Loop counter variable
Symbol Digit = B2               ' Digit is a byte variable
Symbol D1 = B3                  ' Digit 1 pattern
Symbol D2 = B4                  ' Digit 2 pattern
Symbol D3 = B5                  ' Digit 3 pattern
Symbol D4 = B6                  ' Digit 4 pattern
Symbol LED1 = B7                ' Display LED 1 data
Symbol LED2 = B8                ' Display LED 2 data
Symbol Cnt = W6                 ' Cnt is a word variable
Symbol Temp = W7                ' Temp is a word variable
Symbol DATA = Pin6              ' Display Data is RB6
Symbol CLK = 7                  ' Display CLOCK is RB7
```

Figure 5.75 (Continued)

```
'
' START OF MAIN PROGRAM
'
        POKE TRISB, 0                    ' Set PORTB as output

        LED1 = 0                         ' LED 1 is to be off
        LED2 = 0                         ' LED 2 is to be off
        Cnt = 0                          ' Number to display in Cnt

NXT: GOSUB DISPLAY                        ' Display number in Cnt
        PAUSE 1000                       ' Wait 1 second
        Cnt = Cnt + 1                    ' Increment count
        GOTO NXT                         ' Continue counting and displaying


'========================= SUBROUTINES =========================
DISPLAY:
'
' This subroutine displays the number in variable Cnt on the 4-digit 7-segment display
'
        Digit = Cnt / 1000               ' Get 1000s digit
        GOSUB CONVERT                    ' Get segments to turn on
        D4 = Pattern                     ' Pattern for 1000s digit

        Temp = Cnt // 1000               ' Find remainder
        Digit = Temp / 100               ' Get 100s digit
        GOSUB CONVERT                    ' Get segments to turn on
        D3 = Pattern                     ' Pattern for 100s digit

        Temp = Cnt // 100                ' Find remainder
        Digit = Temp / 10                ' Get 10s digit
        GOSUB CONVERT                    ' Get segments to turn on
        D2 = Pattern                     ' Pattern for 10s digit

        Digit = Temp // 10               ' Find remainder
        GOSUB CONVERT                    ' get segments to turn on
        D1 = Pattern                     ' Pattern for 1s digit
'
' Send data to the display. First find out if there are any leading zeroes and
' blank them.
'
        IF D4 = $FC THEN BL4             ' If Digit D4 is zero...
        GOTO CONT                        ' Otherwise continue
```

Figure 5.75   (Continued)

```
BL4:D4 = 0                          ' Blank D4
        IF D3 = $FC THEN BL3        ' If Digit D3 is zero...
        GOTO CONT                   ' Otherwise continue
BL3:D3 = 0                          ' Blank D3
        IF D2 = $FC THEN BL2        ' If Digit D2 is 0...
        GOTO CONT                   ' Otherwise continue
BL2: D2 = 0                         ' Blank D2


CONT:
        GOSUB SEND_START            ' Send START bit
        Pattern = D1
        GOSUB SEGMENTS              ' Send 1s digit
        Pattern = D2
        GOSUB SEGMENTS              ' Send 10s digit
        Pattern = D3
        GOSUB SEGMENTS              ' Send 100s digit
        Pattern = D4
        GOSUB SEGMENTS              ' Send 1000s digit
        GOSUB SEND_LEDS             ' Send LED bits
        GOSUB SEND_TERM             ' Send TERMINATOR bit

CONVERT:
'
' Find the bit pattern to be sent to the display in order to turn on the correct segments
' to display the required number. Digit contains a number between 0 and 9 and
' on return from LOOKUP statement, Pattern contains the bit pattern to send to
' PORTB to display the required number in Digit.
'
        LOOKUP Digit, ($FC, $60, $DA, $F2, $66, $B6, $BE, $E0, $FE, $F6), Pattern
        RETURN


SEND_START:
'
' This subroutine sends a START bit to the display. START bit is a logic 1
'
        DATA = 1                    ' Data = 1
        TOGGLE CLK                  ' CLK = 1
        TOGGLE CLK                  ' CLK = 0
        RETURN


SEND_TERM:
'
' This subroutine sends a TERMINATOR bit to the display. TERMINATOR bit is a logic 0
```

Figure 5.75   (Continued)

```
'
        DATA = 0                    ' Data = 0
        TOGGLE CLK                  ' CLK = 1
        TOGGLE CLK                  ' CLK = 0
        RETURN
```

SEND_LEDS:
```
'
```

' This subroutine sends the two LED data to the display
```
'
        DATA = LED1
        TOGGLE CLK
        TOGGLE CLK
        DATA = LED2
        TOGGLE CLK
        TOGGLE CLK
        RETURN
```

SEGMENTS:
```
'
```

' This subroutine sends data and clock bits to the display. Data bits are sent by left shifting
' the value in variable Pattern. A clock pulse is sent after sending each data bit.
```
'
        FOR I = 1 TO 8
            DATA = Bit7              ' Get bit 7 of Pattern
            TOGGLE CLK               ' CLK = 1
            Pattern = Pattern * 2    ' Shift left pattern 1 digit
            TOGGLE CLK               ' CLK = 0
        NEXT I
        RETURN


        END                         ' End of program
```

Figure 5.75   PicBasic program of Project 16

PicBasic Pro

Figure 5.76 shows the PicBasic Pro program listing which is again smaller and also more efficient than the PicBasic program. Leading zero digits are cleared by checking each leading digit before sending data to it. Leading zero checking is performed as follows:

The bit pattern for all the digit segments are found and if a leading digit is zero and the digit to its left is also zero (variable *First* is 1), then variable *Pattern* is cleared to zero. Byte array "T[ ]" stores the bit patterns of all the digits. A "FOR" loop is formed to shift out the segment data of each digit, with digit 1 bits shifted out first.

```
'***********************************************************************
'
'             4-DIGIT 7-SEGMENT COUNTER WITH BLANKING DISPLAY
'             ========================================================
'
' In this project a B08M04 type 4-digit 7-segment LED displays is used.
' The program counts up by one every second. LED 1 and LED 2 are turned
' off in this example.
'
' The display digits are organised as follows:
'
'        D4      D3      D2      D1
'
' Data is sent: D1 first, then D2, then D3 and finally D4
'
' A PIC16F627 type microcontroller is used in the project with 4MHz
' internal clock and internal reset.
'
' The connection between the display and the microcontroller is as follows:
' (display CE pin is connected to ground permanently)
'
'        RB6     Display DATA
'        RB7     Display CLOCK
'
' In this program the leftmost LEDs which are zero are blanked so that for example
' number 25 is displayed as "25" and not as "0025"
'
'
' Author:        Dogan Ibrahim
' Date:          October, 2005
' Compiler:      PicBasic Pro
' File:          LED29.BAS
'
' Modifications
' =============
'
'***********************************************************************
'
' DEFINITIONS
'
Pattern  VAR Byte              ' Pattern is a byte variable
I        VAR Byte              ' Loop counter variable
Digit    VAR Byte              ' Digit is a byte
```

Figure 5.76 (Continued)

```
First    VAR Byte                    ' Blanking checking variable
T        VAR Byte[4]                 ' Digit segment bit patterns
LED1     VAR Bit                     ' Display LED 1 data
LED2     VAR Bit                     ' Display LED 2 data
Cnt      VAR Word                    ' Cnt is a word variable
Symbol DATA_PIN = PORTB.6           ' Display Data is RB6
Symbol CLK_PIN = PORTB.7            ' Display CLOCK is RB7
'

' START OF MAIN PROGRAM
'

    TRISB = 0                        ' Set PORTB as output

    LED1 = 0                         ' LED 1 is to be off
    LED2 = 0                         ' LED 2 is to be off
    Cnt = 0                          ' Number to display in Cnt

NXT: GOSUB DISPLAY                   ' Display number in Cnt
    PAUSE 1000                       ' Wait 1 second
    Cnt = Cnt + 1                    ' Increment count
    GOTO NXT                         ' Continue counting and displaying


' ===================== SUBROUTINES ===========================
DISPLAY:
'

' This subroutine displays the number in variable Cnt on the 4-digit 7-segment display
'

' Send START bit
'

    DATA_PIN = 1                     ' Data = 1
    PULSOUT CLK_PIN, 1               ' Send a clock
'

' Find out if blanking of leading digits are required or not. Since digit 1 is sent first, we
' have to find all the digits and determine if blanking of any digit is required. Array T[I]
' stores the bit pattern of each digit
'

    First = 1                        ' First time round the loop
    FOR I = 3 TO 0 STEP -1
        Digit = Cnt DIG I            ' Get digits of variable Cnt
        LOOKUP Digit, [$FC, $60, $DA, $F2, $66, $B6, $BE, $E0, $FE, $F6], Pattern
        IF (Digit = 0) AND (First = 1) THEN
                Pattern = 0
        ELSE
                First = 0
        ENDIF
```

Figure 5.76   (Continued)

```
        T[I] = Pattern
    NEXT I

    IF Cnt = 0 THEN T[0] = $FC              ' If Cnt = 0 display 0 in D1
'
' Display each digit with blanking leading zeroes. Digit 1 is sent first
'
    FOR I = 0 To 3
        SHIFTOUT DATA_PIN, CLK_PIN, 1, [ T[I] ]
    NEXT I
'
' Send LED1 and LED 2 bits
'
    DATA_PIN = LED1                         ' Data = LED1
    PULSOUT CLK_PIN,1                       ' Send clock
    DATA_PIN = LED2                         ' Data = LED 2
    PULSOUT CLK_PIN,1                       ' Send clock
'
' Send TERMINATOR bit
'
    DATA_PIN = 0                            ' Data = 0
    PULSOUT CLK_PIN,1                       ' Send clock
    RETURN

    END                                    ' End of program
```

Figure 5.76   PicBasic Pro program of Project 16

Project 17

Project title: 4-digit external interrupt-driven event counter

Project description: This project can be used to count external events and to display the event count on a 4-digit display. An event can be an object on a conveyor belt, number of people entering a building, number of cars entering and leaving a car park, etc. External interrupt input of the microcontroller is used to detect events. An event is detected when the external interrupt pin changes state from logic 1 to logic 0. This project shows how the external interrupt pin of a PIC microcontroller can be used.

Hardware: The circuit diagram of the project is shown in Figure 5.77. Display is connected to bit 6 and bit 7 of PORTB as in Project 16. Interrupt input of the microcontroller (INT) is connected to a switch which simulates the occurrence of an event. The switch is normally at logic 1 and goes to logic 0 when an external event occurs (i.e. when the switch is pressed).

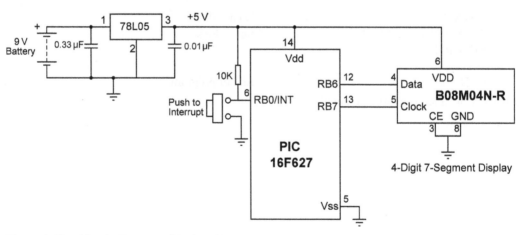

Figure 5.77   Circuit diagram of Project 17

Flow diagram: The flow diagram of the project is given in Figure 5.78. At the beginning of the program event counter variable *Cnt* is cleared and external interrupts are enabled. The main program then goes into an endless loop where the value of *Cnt* is displayed continuously. Whenever an external interrupt occurs the value of event counter *Cnt* is incremented by one and new value of *Cnt* is displayed by the main program.

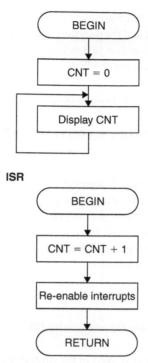

Figure 5.78   Flow diagram of Project 17

Software: **PicBasic**

Interrupts are not directly supported from the PicBasic language and thus only the PicBasic Pro program of this project is given.

PicBasic Pro

Figure 5.79 shows the PicBasic Pro program listing. At the beginning of the program TRISB is set to 1 so that RB0 is configured as input and other bits of PORTB are configured as outputs. Register OPTION_REG is then configured so that external interrupts are recognised on the falling edge (high to low transition) of the interrupt input. Register INTCON is configured to enable external interrupts and the routine starting with label ISR has been assigned to be the interrupt service routine. Notice that the statement "ON INTERRUPT GOTO ISR" assigns label ISR to be the starting address of the interrupt service routine (any other label name can be used here).

Inside the main program variable *Cnt* is cleared and the program calls to subroutine DISPLAY to show the value of variable *Cnt* continuously.

Inside the ISR variable *Cnt* is incremented by 1, external interrupts are re-enabled, and the program returns to the main program. Notice that interrupts are disabled just before entering the ISR, and they are re-enabled just after leaving the ISR.

```
'*********************************************************************
'
'              4-DIGIT INTERRUPT BASED EVENT COUNTER
'              =======================================
'
' In this project a B08M04 type 4-digit 7-segment LED displays is used.
' A switch is connected to the external interrupt input of the microcontroller.
' The program counts external interrupts (i.e. external events) and displays
' the result on a 4-digit 7-segment display. Interrupts are detected on the
' high to low transition of the interrupt pin (RB0/INT) of the microcontroller.
' The display digits are organised as follows:
'
'       D4      D3      D2      D1
'
' Data is sent: D1 first, then D2, then D3 and finally D4
'
' A PIC16F627 type microcontroller is used in the project with 4MHz
' internal clock and internal reset.
'
' The connection between the display and the microcontroller is as follows:
' (display CE pin is connected to ground permanently)
'
'       RB6     Display DATA
'       RB7     Display CLOCK
'
' In this program the leftmost LEDs which are zero are blanked so that for example
' number 25 is displayed as " 25" and not as "0025"
'
'
' Author:       Dogan Ibrahim
' Date:         October, 2005
' Compiler:     PicBasic Pro
' File:         LED30.BAS
'
' Modifications
' =============
'
'*********************************************************************
```

Figure 5.79   (Continued)

```
'
' DEFINITIONS
'

Pattern  VAR Byte                    ' Pattern is a byte variable
I        VAR Byte                    ' Loop counter variable
Digit    VAR Byte                    ' Digit is a byte
First    VAR Byte                    ' Blanking checking variable
T        VAR Byte[4]                 ' Digit segment bit patterns
LED1     VAR Bit                     ' Display LED 1 data
LED2     VAR Bit                     ' Display LED 2 data
Cnt      VAR Word                    ' Cnt is a word variable
Symbol DATA_PIN = PORTB.6            ' Display Data is RB6
Symbol CLK_PIN = PORTB.7            ' Display CLOCK is RB7
'

' START OF MAIN PROGRAM
'
        TRISB = 1                    ' RB0 is input, others output

        ON INTERRUPT GOTO ISR        ' Interrupt service routine
        OPTION_REG = %01000000       ' External interrupt on falling edge of RB0

        LED1 = 0                     ' LED 1 is to be off
        LED2 = 0                     ' LED 2 is to be off
        Cnt = 0                      ' Clear the event counter, Cnt
        INTCON = %10010000           ' Enable external interrupt RB0

NXT: GOSUB DISPLAY                   ' Display number in Cnt
     GOTO NXT                        ' Continue counting and displaying

'
' This is the interrupt service routine, ISR. The program jumps here whenever an external
' interrupt (i.e. whenever an event occurs) occurs
'

DISABLE                              ' Disable interrupts
ISR:                                 ' Entry point of the ISR
    Cnt = Cnt +1                     ' Increment event counter, Cnt
    INTCON = %10010000               ' Enable external interrupts
    RESUME                           ' Resume main program
    ENABLE                           ' Enable interrupts


' ===================== SUBROUTINES =====================
DISPLAY:
'
' This subroutine displays the number in variable Cnt on the 4-digit 7-segment display
```

Figure 5.79   (Continued)

```
'
' Send START bit
'
        DATA_PIN = 1                          ' Data = 1
        PULSOUT CLK_PIN, 1                    ' Send a clock
'
' Find out if blanking of leading digits are required or not. Since digit 1 is sent first, we
' have to find all the digits and determine if blanking of any digit is required. Array T[I]
' stores the bit pattern of each digit
'
        First = 1                             ' First time round the loop
        FOR I = 3 TO 0 STEP -1
            Digit = Cnt DIG I                 ' Get digits of variable Cnt
            LOOKUP Digit, [$FC, $60, $DA, $F2, $66, $B6, $BE, $E0, $FE, $F6], Pattern
            IF (Digit = 0) AND (First = 1) THEN
                    Pattern = 0
            ELSE
                    First = 0
            ENDIF
            T[I] = Pattern
        NEXT I
        IF Cnt = 0 THEN T[0] = $FC            ' If Cnt = 0 display 0 in D1 position
'
' Display each digit with blanking leading zeroes. Digit 1 is sent first
'
        FOR I = 0 To 3
            SHIFTOUT DATA_PIN, CLK_PIN, 1, [ T[I] ]
        NEXT I
'
' Send LED1 and LED 2 bits
'
        DATA_PIN = LED1                       ' Data = LED1
        PULSOUT CLK_PIN,1                     ' Send clock
        DATA_PIN = LED2                       ' Data = LED 2
        PULSOUT CLK_PIN,1                     ' Send clock
'
' Send TERMINATOR bit
'
        DATA_PIN = 0                          ' Data = 0
        PULSOUT CLK_PIN,1                     ' Send clock
        RETURN

        END                                   ' End of program
```

Figure 5.79 PicBasic Pro program of Project 17

Project 18

Project title: 4-digit timer interrupt-driven chronograph

Project description: This project is a chronograph with three push-button switches labelled: START, STOP, and CLEAR. The chronograph is configured to count up accurately in 10 ms intervals using the timer interrupts of the microcontroller. The count is displayed continuously on a 4-digit 7-segment display. Counting starts when the START button is pressed, and stops when the STOP button is pressed. When the counter is in stopped, pressing the CLEAR button clears the display so that a new count can be started.

Hardware: The circuit diagram of the project is shown in Figure 5.80. Display is connected to bit 6 and bit 7 of PORTB as in Project 16. START, STOP, and CLEAR buttons are connected to bit 0, bit 1, and bit 2 of PORTB, respectively. A PIC16F627-type microcontroller is used in this project with 4 MHz internal clock and the internal master clear circuit of the microcontroller is enabled to minimise external component count.

Flow diagram: The flow diagram of the project is given in Figure 5.81. At the beginning of the program event counter variable *Cnt* is cleared and the program waits

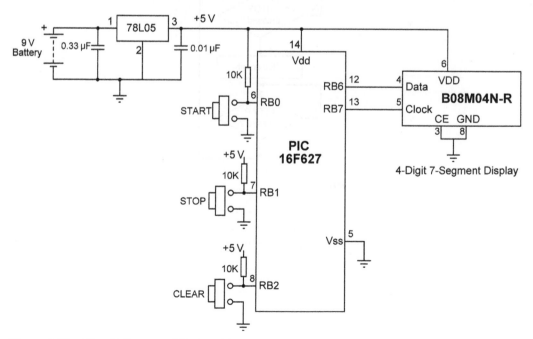

Figure 5.80 Circuit diagram of Project 18

until switch START is pressed. When START is pressed timer interrupt TMR0 is enabled and *Cnt* is incremented every 10 ms. Counting stops when the STOP button is pressed. At this mode, pressing the CLEAR switch clears the display and the process repeats from the beginning.

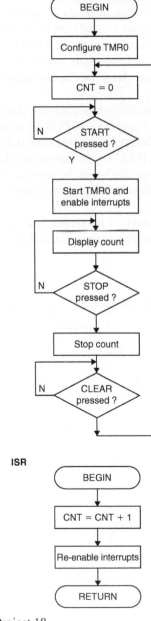

Figure 5.81   Flow diagram of Project 18

Software: **PicBasic**
Interrupts are not directly supported from the PicBasic language and thus
only the PicBasic Pro program of this project is given.

PicBasic Pro
Figure 5.82 shows the PicBasic Pro program listing. At the beginning of
the program TRISB is set to 7 so that RB0, RB1, and RB2 are configured
as inputs and other pins of PORTB are configured as outputs. Then, the
ISR address is defined and OPTION_REG register is set so that the
TMR0 pre-scaler is 64. The program then waits until the START button
(PSTART) is pressed. When the START button is pressed the timer interrupt is enabled so that *Cnt* is incremented automatically every 10 ms. The
program also checks to see if the STOP button (PSTOP) is pressed and if
so, jumps to label STP to stop the timer interrupt. The final value of *Cnt*
can be seen on the display at this point. The BUTTON statement is used
with debouncing to check the state of the push-button switches.

The interrupt service routine starts with label ISR. Inside this routine *Cnt*
is incremented by 1, TMR0 is re-loaded with 100 so that timer interrupts
are generated every 10 ms. Timer interrupt flag (bit 2 of INTCON) is also
cleared inside this routine so that further timer interrupts can be accepted
by the microcontroller. Assuming a 4-MHz clock is used, the timer interrupt TMR0 interval is given by

$$\text{Interval } (\mu s) = \text{pre-scaler} \times (256 - \text{TMR0})$$

with a pre-scaler value of 64 and a TMR0 value of 100 the TMR0 interval is given by

$$\text{Interval} = 64*(256 - 100) = 9.984\,\text{ms}$$

which is close enough to 10 ms … .


```
'********************************************************************
'
'               4-DIGIT TIMER INTERRUPT BASED CHRONOGRAPH
'               ==================================================
'
' In this project a B08M04 type 4-digit 7-segment LED display is used.
' Three switches are connected to the microcontroller: START, STOP and
' CLEAR. When START switch is pressed counting starts with
' 10ms intervals. When the STOP switch is pressed counting stops.
```

Figure 5.82   (Continued)

' When the count stops, pressing the CLEAR button clears the display,
' ready for the next count.
'

' The display digits are organised as follows:
'

'   D4   D3   D2   D1
'

' Data is sent: D1 first, then D2, then D3 and finally D4
'

' A PIC16F627 type microcontroller is used in the project with 4MHz
' internal clock and internal reset.
'

' The connection between the display and the microcontroller is as follows:
' (display CE pin is connected to ground permanently)
'

'   RB6   Display DATA
'   RB7   Display CLOCK
'

' The switches are connected as follows:
'

'   RB0   START
'   RB1   STOP
'   RB2   CLEAR
'

' In this program the leftmost LEDs which are zero are blanked so that for example
' number 25 is displayed as "25" and not as "0025"
'

'

' Author: Dogan Ibrahim
' Date: October, 2005
' Compiler: PicBasic Pro
' File: LED31.BAS
'

' Modifications
' ==========
'

'**

'

' DEFINITIONS
'

Pattern VAR Byte   ' Pattern is a byte variable
I   VAR Byte   ' Loop counter variable

Figure 5.82   (Continued)

```
Digit    VAR Byte                          ' Digit is a byte
First    VAR Byte                          ' Blanking checking variable
temp     VAR Byte                          ' Temporary byte variable
T        VAR Byte[4]                       ' Digit segment bit patterns
LED1     VAR Bit                           ' Display LED 1 data
LED2     VAR Bit                           ' Display LED 2 data
Cnt      VAR Word                          ' Cnt is a word variable
Symbol DATA_PIN = PORTB.6                  ' Display Data is RB6
Symbol CLK_PIN = PORTB.7                   ' Display CLOCK is RB7
Symbol PSTART = PORTB.0                    ' START button
Symbol PSTOP = PORTB.1                     ' STOP button
Symbol PCLEAR = PORTB.2                    ' CLEAR button

'
' START OF MAIN PROGRAM
'
     TRISB = 7                             ' RB0, RB1, RB2 are inputs, others output

     ON INTERRUPT GOTO ISR                 ' Interrupt service routine
     OPTION_REG = %00000101                ' Configure TMR0 for prescaler=64

     LED1 = 0                              ' LED 1 is to be off
     LED2 = 0                              ' LED 2 is to be off
BEGIN:
     Cnt = 0                               ' Clear the event counter, Cnt
     GOSUB DISPLAY                         ' Display Cnt
     INTCON = %10010000                    ' Enable timer interrupt TMR0
'
' Wait until the START button is pressed
'
BT: temp = 0
     BUTTON PSTART, 0, 255, 0, temp, 1, STRT    ' Goto STRT if START pressed
     GOTO BT
'
' START button is pressed. Start the counting
'
STRT:
     TMR0 = 100                            ' count = 64*(256 − 100) = 9984us
     INTCON = %10100000                    ' Enable TMR0 interrupts
WT:  GOSUB DISPLAY                         ' Display Cnt
     IF PSTOP = 0 THEN STP                 ' If STOP switch is pressed
     GOTO WT                               ' Repeat
```

Figure 5.82   (Continued)

```
'
' STOP button is pressed. Stop the counting
'
STP:
      INTCON = 0                                      ' Stop counting
 ST: temp = 0
      BUTTON PCLEAR, 0, 255, 0, temp, 1, BEGIN
      GOTO ST
'
' This is the timer interrupt service routine, ISR. The program jumps here whenever a timer
' interrupt occurs (i.e. every 10ms)
'

DISABLE                                              ' Disable interrupts
ISR:                                                 ' Entry point of the ISR
      TMR0 = 100                                     ' TMR0 value for 10ms count
      Cnt = Cnt +1                                   ' Increment event counter, Cnt
      INTCON.2 = 0                                   ' Re-enable TMR0 interrupts
      RESUME                                         ' Resume main program
      ENABLE                                         ' Enable interrupts



' =========================== SUBROUTINES ===========================
DISPLAY:
'
' This subroutine displays the number in variable Cnt on the 4-digit 7-segment display
'
' Send START bit
'
      DATA_PIN = 1                                   ' Data = 1
      PULSOUT CLK_PIN, 1                             ' Send a clock
'
' Find out if blanking of leading digits are required or not. Since digit 1 is sent first, we
' have to find all the digits and determine if blanking of any digit is required. Array T[I]
' stores the bit pattern of each digit
'
      First = 1                                      ' First time round the loop
      FOR I = 3 TO 0 STEP -1
         Digit = Cnt DIG I                           ' Get digits of variable Cnt
         LOOKUP Digit, [$FC, $60, $DA, $F2, $66, $B6, $BE, $E0, $FE, $F6], Pattern
         IF (Digit = 0) AND (First = 1) THEN
               Pattern = 0
```

Figure 5.82   (Continued)

```
        ELSE
                First = 0
        ENDIF
        T[I] = Pattern
    NEXT I

    IF Cnt = 0 THEN T[0] = $FC              ' If Cnt = 0 then display 0 in D1 position
'
' Display each digit with blanking leading zeroes. Digit 1 is sent first
'
    FOR I = 0 To 3
    SHIFTOUT DATA_PIN, CLK_PIN, 1, [ T[I] ]
    NEXT I
'
' Send LED1 and LED 2 bits
'
    DATA_PIN = LED1                         ' Data = LED1
    PULSOUT CLK_PIN,1                       ' Send clock
    DATA_PIN = LED2                         ' Data = LED 2
    PULSOUT CLK_PIN,1                       ' Send clock
'
' Send TERMINATOR bit
'
    DATA_PIN = 0                            ' Data = 0
    PULSOUT CLK_PIN,1                       ' Send clock
    RETURN

    END                                    ' End of program
```

Figure 5.82   PicBasic Pro program of Project 18

Project 19

Project title: Car park control system

Project description: This project is a simple car park control system. Two barriers are used, one at the entry and one at the exit of a car park. When a barrier is lifted to allow a car to pass through, switches are activated which send logic 0 pulses to the microcontroller. A 4-digit 7-segment display is connected to the output of the control system. The system counts the difference of the number of cars entering and leaving the car park. If the count is less than 100 (assuming the car park can take up to 100 cars) the message SPCS (i.e. spaces) will be displayed. When the car park is full, the message FULL will be displayed. Assume that the barriers lift-up automatically when a vehicle approaches them. Also assume that the entry barrier has a locking mechanism and this mechanism is enabled to lock the barrier so that it does not lift-up when the car park is full.

Figure 5.83 shows the block diagram of the car park control system.

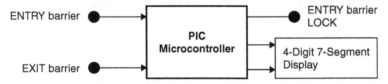

Figure 5.83   Block diagram of Project 19

Hardware: The circuit diagram of the project is shown in Figure 5.84. Display is connected to bit 6 and bit 7 of PORTB as in Project 16. ENTRY and EXIT switches are connected to bit 0 and bit 1 of PORTB, respectively. ENTRY barrier lock output is connected to bit 2 of PORTB. In Figure 5.84, the barrier switches are shown as simple push-button switches. Also, the ENTRY lock mechanism is an output from RB2 and is shown as a small circle.

In this project a PIC16F627-type microcontroller is used and the microcontroller is operated with its 4 MHz internal clock and internal master clear circuit.

Flow diagram: The flow diagram of the project is given in Figure 5.85. At the beginning of the program event counter variable *Cnt* is cleared and the program checks the value of *Cnt*. If *Cnt* $\geq$ *100* then the car park is assumed to be full and message "FULL" is displayed. If on the other hand *Cnt* $<$ *100* then it is assumed that there are spaces in the car park and message

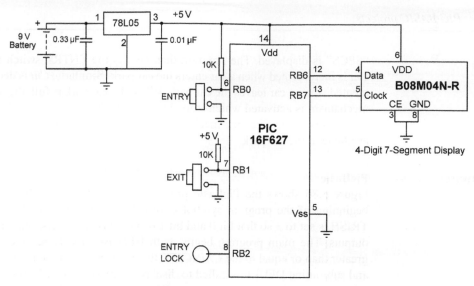

Figure 5.84 Circuit diagram of Project 19

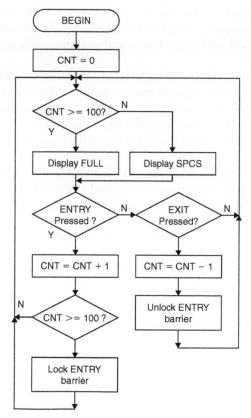

Figure 5.85 Flow diagram of Project 19

"SPCS" is displayed. The program then checks the ENTRY switch and *Cnt* is incremented when a car enters the car park. Similarly, *Cnt* is decremented when a car leaves the car park. When the car park is full, the lock mechanism is activated which stops The ENTRY barrier to open when a car approaches it. The lock mechanism is disabled as soon as spaces are available in the car park.

Software:

PicBasic

Figure 5.86 shows the PicBasic program listing of Project 19. At the beginning of the program symbol CAPACITY is assigned to 100 and TRISB is set to 3 so that bit 0 and bit 1 of PORTB are inputs, other pins outputs. The main program begins with label BEGIN. Here, if *Cnt* is greater than or equal to the *CAPACITY*, the car park is assumed to be full and subroutine DFULL is called to display the message FULL. If on the other hand *Cnt* is less than the *CAPACITY* then the car park is assumed to have spaces and subroutine DSPCS is called to display the message SPCS. ENTRY and EXIT switches are checked inside the LOOP. When a vehicle enters the car park, ENTRY switch is activated and program jumps to label LENTRY. Similarly, when a vehicle leaves the car park, EXIT switch is activated and program jumps to label LEXIT.

Inside subroutine LENTRY, *Cnt* is incremented by 1 and LOCK is set to 1 if *Cnt* is greater than or equal to the *CAPACITY*. The program then jumps to label BEGIN to repeat the process. Inside the LEXIT subroutine, *Cnt* is decremented By 1 and LOCK is cleared. The program then jumps to label BEGIN to repeat the process.

Characters FULL and SPCS are obtained by loading D1 – D4 with the correct bit patterns for these characters. Table 5.9 shows how to obtain characters FULL and SPCS by sending hexadecimal data to the display.

Thus, to FULL will be displayed if the following hexadecimal numbers are sent to the display:

$1C $1C $7C $8E

Similarly, SPCS will be displayed if the following hexadecimal numbers are sent to the display:

$B6 $9C $CE $B6

Bit patterns are sent to the display using subroutine SHIFTO which sends out data bits in serial form with clock.

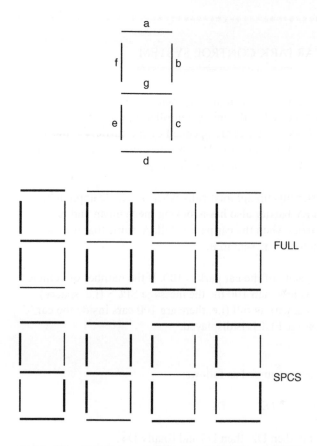

Table 5.9 Bit patterns for characters FULL and SPCS

Character	a b c d e f g dp	Hexadecimal
F	1 0 0 0 1 1 1 0	$8E
U	0 1 1 1 1 1 0 0	$7C
L	0 0 0 1 1 1 0 0	$1C
S	1 0 1 1 0 1 1 0	$B6
P	1 1 0 0 1 1 1 0	$CE
C	1 0 0 1 1 1 0 0	$9C

```
'*************************************************************
'
'                 CAR PARK CONTROL SYSTEM
'                 ===========================
'
' In this project a B08M04 type 4-digit 7-segment LED display is used.
' Two switches are connected to the microcontroller inputs: ENTRY switch
' and EXIT switch. These switches are operated by the barriers at the
' entrance and the exit of the car park. The switches are normally at logic
' 1 and they go to logic 0 when a barrier is lifted up.
'
' Assume that a barrier lifts up automatically when a vehicle approaches
' a barrier. The ENTRY barrier also has a locking mechanism and is
' used to lock the barrier when the car park is full. Assume that the lock
' is activated when a logic 1 is sent to it.
'
' Assume that the capacity of the car park is 100. If the number of vehicles
' inside the car park is less than 100 the the message SPCS (i.e. spaces)
' is displayed. If the car park is full (i.e. there are 100 cars inside the car
' park), then the message FULL is displayed.
'
'
' The display digits are organised as follows:
'
'         D4      D3      D2      D1
'
' Data is sent: D1 first, then D2, then D3 and finally D4
'
' A PIC16F627 type microcontroller is used in the project with 4MHz
' internal clock and internal reset.
'
' The connection between the display and the microcontroller is as follows:
' (display CE pin is connected to ground permanently)
'
'         RB6     Display DATA
'         RB7     Display CLOCK
'
' The switches are connected as follows:
'
'         RB0     ENTRY
'         RB1     EXIT
'         RB2     LOCK
```

Figure 5.86   (Continued)

```
'
'
'
' Author:        Dogan Ibrahim
' Date:          October, 2005
' Compiler:      PicBasic
' File:          CAR_PRK1.BAS
'
' Modifications
' ============
'
'*********************************************************************
'
' DEFINITIONS
'
Symbol TRISB = $86                          ' TRISB address
Symbol PORTB = $06                          ' PORTB address
'
Symbol Pattern = B0
Symbol I = B1                               ' Loop counter variable
Symbol temp = B2                            ' Temporary byte variable
Symbol LED1 = B3                            ' Display LED 1 data
Symbol LED2 = B4                            ' Display LED 2 data
Symbol D1 = B5                              ' Digit 1 data
Symbol D2 = B6                              ' Digit 2 data
Symbol D3 = B7                              ' Digit 3 data
Symbol D4 = B8                              ' Digit 4 data
Symbol Cnt = W5
Symbol DATA = Pin6                          ' Display Data is RB6
Symbol CLK = 7                              ' Display CLOCK is RB7
Symbol PENTRY = 0                           ' START button
Symbol PEXIT = 1                            ' STOP button
Symbol LOCK = 2                             ' LOCK output

Symbol CAPACITY = 100                       ' Car park capacity is 100 vehicles
'
' START OF MAIN PROGRAM
'
        POKE TRISB, 3                       ' RB0, RB1 are inputs, others output

        LED1 = 0                            ' LED 1 is to be off
        LED2 = 0                            ' LED 2 is to be off
        Cnt = 0                             ' Clear the car park count
BEGIN:
        IF Cnt >= CAPACITY THEN LARGER
        IF Cnt < CAPACITY THEN SMALLER
```

Figure 5.86   (Continued)

```
LARGER: GOSUB DFULL
        GOTO LOOP
SMALLER: GOSUB DSPCS
'
' Check if ENTRY barrier is lifted up
'

LOOP:
        temp = 0
        BUTTON PENTRY, 0, 255, 0, temp, 1, LENTRY     ' Goto LENTRY if
                                                      ' ENTRY switch = 0

        temp = 0

        BUTTON PEXIT, 0, 255, 0, temp, 1, LEXIT       ' Goto LEXIT if EXIT switch = 0
        GOTO LOOP
'
' START button is pressed. Start the counting
'

LENTRY:
        Cnt = Cnt + 1
        IF Cnt < CAPACITY THEN BEGIN                  ' A vehicle entered the car park
        HIGH LOCK                                     ' Lock the ENTRY barrier
        GOTO BEGIN

LEXIT:
        Cnt = Cnt − 1                                 ' A vehicle left the car park
        LOW LOCK                                      ' Unlock the ENTRY barrier
        GOTO BEGIN

DFULL:
        D1 = $1C:      D2 = $1C:      D3 = $7C:      D4 = $8E
        GOSUB DISPLAY
        RETURN

DSPCS:
        D1 = $B6:      D2 = $9C:      D3 = $CE:      D4 = $B6
        GOSUB DISPLAY
        RETURN

DISPLAY:
'
' This subroutine displays the the 4 byte data in D1,D2,D3,D4.
' D1 data is sent first to the display.
'
' Send start bit
```

Figure 5.86   (Continued)

```
'
'
            DATA = 1                              ' Data = 1
            TOGGLE CLK                            ' CLK = 1
            TOGGLE CLK                            ' CLK = 0
'
' Send segment data
'
            B0 = D1
            GOSUB SHIFTO                          ' Display D1
            B0 = D2
            GOSUB SHIFTO                          ' Display D2
            B0 = D3
            GOSUB SHIFTO                          ' Display D3
            B0 = D4
            GOSUB SHIFTO                          ' Display D4
'
' This subroutine sends the two LED data to the display
'
            DATA = LED1
            TOGGLE CLK
            TOGGLE CLK
            DATA = LED2
            TOGGLE CLK
            TOGGLE CLK
'
' Send terminator bit
'
            DATA = 0                              ' Data = 0
            TOGGLE CLK                            ' CLK = 1
            TOGGLE CLK                            ' CLK = 0
            RETURN

SHIFTO:
'
' This subroutine shifts out data with clock
'
            FOR I = 1 TO 8
                DATA = Bit7                       ' Get bit 7 of Pattern
                TOGGLE CLK                        ' CLK = 1
                Pattern = Pattern * 2             ' Shift left pattern 1 digit
                TOGGLE CLK                        ' CLK = 0
            NEXT I
            RETURN

            END                                   ' End of program
```

Figure 5.86   PicBasic program of Project 19

PicBasic Pro

Figure 5.87 shows the PicBasic Pro program listing. At the beginning of the program, the capacity of the car park, symbol *CAPACITY* is assigned value 100. TRISB register is set to 3 so that bit 0 and bit 1 of PORTB are inputs, the other pins outputs. The main program loop begins with label BEGIN. Here, if *Cnt* is greater than or equal to the *CAPACITY*, the car park is assumed to be full and subroutine DFULL is called to display the message FULL. If on the other hand *Cnt* is less than the *CAPACITY* then the car park is assumed to have spaces and subroutine DSPCS is called to display the message SPCS. The ENTRY and EXIT switches are checked inside the LOOP. When a vehicle enters the car park, ENTRY switch is activated and program jumps to label LENTRY. Similarly, when a vehicle leaves the car park, EXIT switch is activated and program jumps to label LEXIT.

Inside subroutine LENTRY, *Cnt* is incremented by 1 and LOCK is set to 1 if *Cnt* is greater than or equal to the *CAPACITY*. The program then jumps to label BEGIN to repeat the process. Inside the LEXIT subroutine, *Cnt* is decremented By 1 and LOCK is cleared. The program then jumps to label BEGIN to repeat the process. Data bits are sent out using the SHIFTOUT command of PicBasic Pro.

```
'*********************************************************************
'
'               CAR PARK CONTROL SYSTEM
'               ============================
'
'
' In this project a B08M04 type 4-digit 7-segment LED display is used.
' Two switches are connected to the microcontroller inputs: ENTRY switch
' and EXIT switch. These switches are operated by the barriers at the
' entrance and the exit of the car park. The switches are normally at logic
' 1 and they go to logic 0 when a barrier is lifted up.
'
' Assume that a barrier lifts up automatically when a vehicle approaches
' a barrier. The ENTRY barrier also has a locking mechanism and is
' used to lock the barrier when the car park is full. Assume that the lock
' is activated when a logic 1 is sent to it.
'
' Assume that the capacity of the car park is 100. If the number of vehicles
' inside the car park is less than 100 the the message SPCS (i.e. spaces)
' is displayed. If the car park is full (i.e. there are 100 cars inside the car
' park), then the message FULL is displayed.
```

Figure 5.87 (Continued)

```
'
'
' The display digits are organised as follows:
'
'        D4      D3      D2      D1
'
' Data is sent: D1 first, then D2, then D3 and finally D4
'
' A PIC16F627 type microcontroller is used in the project with 4MHz
' internal clock and internal reset.
'
' The connection between the display and the microcontroller is as follows:
' (display CE pin is connected to ground permanently)
'
'        RB6     Display DATA
'        RB7     Display CLOCK
'
' The switches are connected as follows:
'
'        RB0     ENTRY
'        RB1     EXIT
'        RB2     LOCK
'
'
'
' Author:        Dogan Ibrahim
' Date:          October, 2005
' Compiler:      PicBasic Pro
' File:          CAR_PRK2.BAS
'
' Modifications
' ============
'
'
'**************************************************************************
'
' DEFINITIONS
'
I      VAR Byte              ' Loop counter variable
temp   VAR Byte              ' Temporary byte variable
T      VAR Byte[4]           ' Digit segment bit patterns
LED1   VAR Bit               ' Display LED 1 data
LED2   VAR Bit               ' Display LED 2 data
Cnt    VAR Word              ' Cnt is a word variable
```

Figure 5.87   (Continued)

```
Symbol DATA_PIN = PORTB.6          ' Display Data is RB6
Symbol CLK_PIN = PORTB.7           ' Display CLOCK is RB7
Symbol PENTRY = PORTB.0            ' START button
Symbol PEXIT = PORTB.1             ' STOP button
Symbol LOCK = PORTB.2              ' LOCK output

Symbol CAPACITY = 100              ' Car park capacity is 100 vehicles
'
' START OF MAIN PROGRAM
'
        TRISB = 3                  ' RB0, RB1 are inputs, others output

        LED1 = 0                   ' LED 1 is to be off
        LED2 = 0                   ' LED 2 is to be off
        Cnt = 0                    ' Clear the car park count

BEGIN:
        IF Cnt >= CAPACITY THEN
            GOSUB DFULL
        ELSE
            GOSUB DSPCS
        ENDIF


'
' Check if ENTRY barrier is lifted up
'
LOOP:
        temp = 0
        BUTTON PENTRY, 0, 255, 0, temp, 1, LENTRY   ' Goto LENTRY if ENTRY = 0
        temp = 0

        BUTTON PEXIT, 0, 255, 0, temp, 1, LEXIT     ' Goto LEXIT if EXIT switch = 0
        GOTO LOOP
'
' START button is pressed. Start the counting
'
LENTRY:
        Cnt = Cnt + 1                       ' A vehicle entered the car park
        IF Cnt >= CAPACITY THEN LOCK = 1    ' Lock the ENTRY barrier
        GOTO BEGIN

LEXIT:
        Cnt = Cnt - 1                       ' A vehicle left the car park
        LOCK = 0                            ' Unlock the ENTRY barrier
```

Figure 5.87   (Continued)

```
            GOTO BEGIN
DFULL:
            T[0] = $1C:     T[1] = $1C:     T[2] = $7C:     T[3] = $8E
            GOSUB DISPLAY
            RETURN


DSPCS:
            T[0] = $B6:     T[1] = $9C:     T[2] = $CE:     T[3] = $B6
            GOSUB DISPLAY
            RETURN



' This subroutine displays the 4 byte data in array T[I]. T[0] is the
'  first data sent to the display.
'

DISPLAY:
'

' Send START bit
'

            DATA_PIN = 1                        ' Data = 1
            PULSOUT CLK_PIN, 1                  ' Send a clock

'
' Display each digit. Digit 1 is sent first
'

            FOR I = 0 To 3
                SHIFTOUT DATA_PIN, CLK_PIN, 1, [ T[I] ]
                NEXT I


'
' Send LED1 and LED 2 bits
'

            DATA_PIN = LED1                     ' Data = LED1
            PULSOUT CLK_PIN,1                   ' Send clock
            DATA_PIN = LED2                     ' Data = LED 2
            PULSOUT CLK_PIN,1                   ' Send clock


'
' Send TERMINATOR bit
'


            DATA_PIN = 0                        ' Data = 0
            PULSOUT CLK_PIN,1                   ' Send clock
            RETURN


            END                                 ' End of program
```

Figure 5.87 PicBasic Pro program of Project 19

Project 20

Project title: Seconds counter with LCD display

Project description: In this project a seconds counter is used and the count is displayed on a LCD display as follows:

$$CNT = nnn$$

Figure 5.88 shows the block diagram of the project. A PIC microcontroller is used with its outputs connected to a parallel LCD.

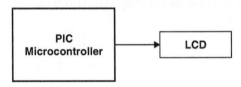

Figure 5.88   Block diagram of Project 20

Hardware: The circuit diagram of the project is shown in Figure 5.89. A PIC16F627 microcontroller is used in the project with a 4 MHz internal clock and the internal master clear circuit enabled during programming of the chip. In this project a 2 row LCD is used but any type LCD can be used as long as it is compatible with the HD44780 chip. The LCD display is connected to the microcontroller using the default connections described in Section 4.3, i.e. the following connections are made between the microcontroller and the LCD display:

LCD	Microcontroller
D4	RA0
D5	RA1
D6	RA2
D7	RA3
E	RB3
RS	RA4

Notice that pin RA4 of the microcontroller is open-drain output and should be connected to the +V supply with a 10 K resistor.

The project constructed on a breadboard is shown in Figure 5.90.

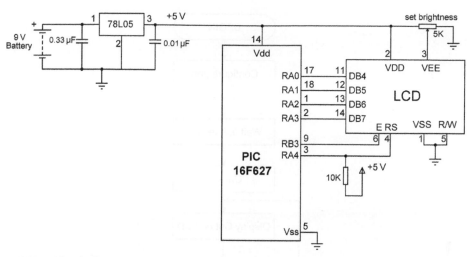

Figure 5.89 Circuit diagram of Project 20

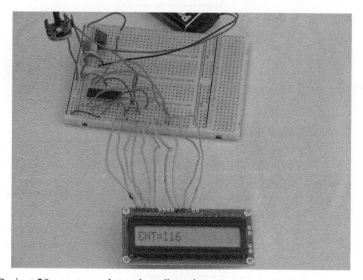

Figure 5.90 Project 20 constructed on a breadboard

Flow diagram: The flow diagram of the project is given in Figure 5.91. At the beginning of the program PORTA and PORTB directions are configured. The program then waits for about 0.5 s for the LCD to initialise. Variable *Cnt* is incremented every second and the result is displayed on the LCD in the following format:

$$CNT = nnn$$

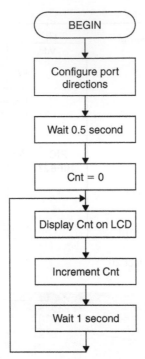

Figure 5.91   Flow diagram of Project 20

Software:　　　　　　　**PicBasic Pro**

Figure 5.92 shows the PicBasic Pro program listing of the project. At the beginning of the program PORTA and PORTB are configured as outputs. Register CMCO is set to 7 so that PORTA pins are configured as digital I/O. *Cnt* is declared as a word variable and program waits 500 ms for the initialisation of the LCD. The LCD is cleared and the cursor is set to the home position using the $FE,1 and $FE,2 LCD commands, respectively. Variable *Cnt* is then displayed on the LCD in decimal format using the LCDOUT statement. The program waits for 1 s and the process is repeated after *Cnt* is incremented by one.

```
'*********************************************************************
'
'                  LCD SECONDS COUNTER
'                  ====================
'
' In this project an LCD display is connected to a PIC16F627 microcontroller.
' The microcontroller is configured to operate with a 4MHz internal clock.
```

Figure 5.92   (Continued)

```
'
' Variable Cnt is incremented by 1 every second and the result is displayed on
' the LCD.
'
' The connection between the LCD display and the microcontroller is as follows:
'
'        Display          Microcontroller pin
'        DB4              RA0
'        DB5              RA1
'        DB6              RA2
'        RB7              RA3
'        E                RB3
'        RS               RA4
'
'
' A 10K resistor is used to pull-up pin RA4 of the microcontroller.
'
' RW pin of the LCD is connected to ground. The brightness of the LCD is
' controlled by connecting a 5K variable resistor to pin VEE of the display.
'
'
' Author:      Dogan Ibrahim
' Date:        November, 2005
' Compiler:    PicBasic Pro
' File:        LCD1.BAS
'
' Modifications
' ===========
'
'*********************************************************************
'
' DEFINITIONS
'
Cnt     VAR Word                         ' Cnt is a word variable
'
' START OF MAIN PROGRAM
'
      CMCON = 7                          ' RA0-RA3 are digital I/O
      TRISA = 0                          ' PORTA is output
      TRISB = 0                          ' PORTB is output

      PAUSE 500                          ' Wait 0.5 second to initialize LCD
      Cnt = 0                            ' Clear Cnt to zero
      LCDOUT $FE,1                       ' Clear LCD
```

Figure 5.92   (Continued)

RPT:

LCDOUT $FE,2	' Home cursor
LCDOUT "CNT = ", DEC Cnt	' Display count
PAUSE 1000	' Wait 1 second
Cnt = Cnt +1	' Increment Cnt
GOTO RPT	' Repeat
END	' End of program

Figure 5.92   PicBasic Pro program of Project 20

PicBasic

PicBasic language does not provide any instructions to drive an LCD directly. It is however possible to develop low-level routines to initialise and drive LCD displays. The details of these routines require a detailed knowledge of the internal operations of the LCDs and are only described here briefly.

The steps to initialise an LCD are given below. Here, we are assuming that the LCD is connected to the microcontroller in the standard way as shown in Figure 5.89.

- Wait 20 ms after power up
- DO 3 times
  - Send 3 to LCD
  - Wait 10 ms
  - Toggle Enable line
- ENDDO
- Wait 10 ms
- Send 2 to LCD (LCD in 4-bit mode)
- Wait 1 ms
- Toggle Enable line
- Send $28 to LCD (4-bit, 2-lines, 5 × 7 font)
- Send $0C to LCD (display on, no cursor, no blink)
- Send $06 to LCD (LCD entry mode, no shift)

Figure 5.93 shows the PicBasic program listing of the project. The majority of the code used is to initialise and drive the LCD. At the beginning of the program the addresses of the SFR registers used in the program are defined. Then, PORTA and PORTB ports are configured as outputs. The program then jumps to subroutine INITLCD to initialise the LCD. The initialisation is basically in three steps: *resetting mode of* the LCD, *function setting* of the LCD, *display on* routine, and the *entry mode*. The LCD can be operated in either 4-bit or 8-bit modes. Operating in 4-bit mode has the advantage that only 4 data lines and 2 control lines, i.e. a total of 6 lines are required to initialise and control the LCD. At the beginning of the initialisation routine the program waits for 20 ms for the internal logic of the LCD to be initialised. At this point the LCD is by default in 8-bit mode. Then the *resetting mode* starts where a data byte 3 should be sent to the LCD data lines with a delay between each data output, and the Enable line of the LCD should be toggled after each output. The recommended delay is at least 4.1 ms after the first output and at least 100 μs after the other two outputs. In this example, a 10 ms delay is used after each output and the PULSOUT statement is used to toggle the Enable line (bit 3 of PORTB – RB3) of the LCD.

LCD is then in *function setting* mode where the LCD is put into the 4-bit mode and the character font is selected. In the *display on* mode the display and the cursor are turned on. The final stage of the initialisation is the *entry mode* where the cursor movement mode and cursor blinking are specified.

Subroutine LCDDATA can be used to display the character in register B2 on the current cursor position. The high nibble is first obtained by shifting the data right by 4 bits. This nibble is sent to the LCD and the Enable line is toggled. The low nibble is then sent to the LCD. LCD pin RS is set to logic 1 during the data mode.

Subroutine SENDCOM is used to send a command to the LCD. LCD pin RS is cleared to logic 0 during the command mode. The command in register B2 is sent to the LCD. The Enable line is toggled after sending each nibble of the command.

Subroutine CLRLCD is used to clear the LCD display. Similarly, subroutine HOMELCD can be used to set the cursor to the home position. Other LCD commands (e.g. to move the cursor left or right, to move to the second line, etc.) can easily be added to the program.

Variable *Cnt* is declared as a word and initialised to zero. The FOR loop after label RPT uses statement LOOKUP to extract the characters of the string to be displayed ("CNT=" in this case) where the characters are stored in register B2 and then displayed by calling subroutine LCDDATA.

The 100 s, 10 s, and 1 s digits of variable *Cnt* are then extracted and stored in registers B4, B5, and B6, respectively. For example, if Cnt = 573, then B4 = 5, B5 = 7, and B6 = 3. Leading zeroes are suppressed by not displaying them. The extracted numbers are then converted into ASCII by adding 48 (ASCII "0") to each digit. The digits are assigned to register B2 and displayed on the LCD by calling subroutine LCDDATA. The maximum value of *Cnt* that can be displayed is 999 (this value can be increased by extracting the 1000s digit of *Cnt*).

The program then waits for 1 s, variable *Cnt* is incremented by one and the program repeats.

```
'*******************************************************************
'
'                 LCD SECONDS COUNTER
'                 ========================
'
' In this project an LCD display is connected to a PIC16F627 microcontroller.
' The microcontroller is configured to operate with a 4MHz internal clock.
'
' Variable Cnt is incremented by 1 every second and the result is displayed on
' the LCD.
'
' The connection between the LCD display and the microcontroller is as follows:
'
'         Display         Microcontroller pin
'         DB4             RA0
'         DB5             RA1
'         DB6             RA2
'         RB7             RA3
'         E               RB3
'         RS              RA4
'
'
' A 10K resistor is used to pull-up pin RA4 of the microcontroller.
'
' RW pin of the LCD is connected to ground. The brightness of the LCD is
' controlled by connecting a 5K variable resistor to pin VEE of the display.
'
' LCD is initialized and controlled by using low-level LCD routines.
'
'
' Author:      Dogan Ibrahim
' Date:        November, 2005
' Compiler:    PicBasic
' File:        LCD2.BAS
```

Figure 5.93   (Continued)

```
'
' Modifications
' ===========
'
'*********************************************************************

'
' DEFINITIONS
'
Symbol PORTA = 5                        ' PORTA address
Symbol TRISA = $85                      ' TRISA address
Symbol PORTB = 6                        ' PORTB address
Symbol CMCON = $1F                      ' CMCON address
Symbol TRISB = $86                      ' PORTB address
Symbol Cnt = W0                         ' Cnt is a word variable


' START OF MAIN PROGRAM
'
        POKE CMCON, 7                   ' RA0-RA3 are digital I/O
        POKE TRISA, 0                   ' PORTA is output
        POKE TRISB, 0                   ' PORTB is output

        GOSUB INITLCD                   ' Initialize LCD

        Cnt = 0                         ' Clear Cnt to zero
        GOSUB CLRLCD                    ' Clear LCD
RPT:
        GOSUB HOMELCD                   ' Home cursor

        FOR B4 = 0 TO 3
            LOOKUP B4, ("CNT = "), B2
            GOSUB LCDDATA               ' Display CNT =
        NEXT B4
'
' Find the 100s, 10s, and 10s digits and store in registers B4, B5, B6 respectively.
' For example, if the number (Cnt) is 234, then B4 = 2, B5 = 3 and B6 = 4. Numbers
' up to 999 can be displayed. i.e. maximum value of Cnt is 999. The program also blanks
' zeroes from the beginning. e.g. if Cnt = 5, then only 5 is displayed. i.e.
' 005 is not displayed.
'
        B4 = Cnt / 100                  ' B4 = leftmost digit
        B6 = Cnt // 100
        B5 = B6 / 10                    ' B5 = middle digit
```

Figure 5.93   (Continued)

```
    B6 = B6 // 10                      ' B6 = rightmost digit
    IF B4 = 0 THEN NO1
    B2 = B4 + 48
    GOSUB LCDDATA                      ' Display top digit
CONT:
    B2 = B5 + 48
    GOSUB LCDDATA                      ' Display middle digit
    B2 = B6 + 48
    GOSUB LCDDATA                      ' Display rightmost digit
    GOTO NXT
NO1:
    IF B5 = 0 THEN NO2
    GOTO CONT
NO2:
    B2 = B6 + 48
    GOSUB LCDDATA
'
' Wait 1 second, increment Cnt and repeat
'
NXT: PAUSE 1000                        ' Wait 1 second
    Cnt = Cnt +1                       ' Increment Cnt
    GOTO RPT                           ' Repeat


'
'           SUBROUTINES
'           =============
'
' This subroutine initializes the LCD
'
INITLCD:
    PAUSE 20                           ' Wait 20ms

    FOR B2 = 1 TO 3                    ' Do 3 times
        POKE PORTA, 3
        PULSOUT 3, 100                 ' Toggle Enable line
        PAUSE 10                       ' Wait 10ms
    NEXT B2

    PAUSE 10                           ' Wait 10ms

    POKE PORTA, 2                      ' Send 2 to LCD
    PULSOUT 3, 100                     ' Toggle Enable line

    B2 = $28                           ' Send $28 to LCD
    GOSUB SENDCOM
```

Figure 5.93   (Continued)

```
    B2 = $0C                          ' Send $0C to LCD
    GOSUB SENDCOM

    B2 = $06                          ' Send $06 to LCD
    GOSUB SENDCOM
    RETURN


'
' This subroutine clears the LCD
'
CLRLCD:
    B2 = 1
    GOSUB SENDCOM
    PAUSE 2
    RETURN


'
' This subroutine homes the cursor
'
HOMELCD:
    B2 = 2
    GOSUB SENDCOM
    PAUSE 5
    RETURN
'
' This subroutine sends data to the LCD. Data to be output is assumed to be
' in register B2.
'
LCDDATA:
    B3 = B2 / 16                      ' Shift B2 right 4 times
    B3 = B3 + 16                      ' Add the LCD RS bit
    POKE PORTA, B3                    ' Send to LCD
    PULSOUT 3, 100                    ' Toggle Enable line

    B3 = B2 & $0F                     ' Extract 4 low order bits
    B3 = B3 + 16                      ' Add the LCD RS bit
    POKE PORTA, B3                    ' Send to LCD
    PULSOUT 3, 100                    ' Toggle Enable line
    PAUSE 2                           ' Wait 2ms to complete
    RETURN
```

Figure 5.93   (Continued)

'

' This subroutine sends a command to the LCD. The comamnd is in B2.
' We have to shift top 4 bits down to the bottom 4 bits.

'

```
SENDCOM:
        B3 = B2 / 16                    ' Shift B2 right 4 times
        B3 = B3 & $EF                   ' Clear RS = 0
        POKE PORTA, B3                  ' Send B3 to LCD
        PULSOUT 3, 100                  ' Toggle Enable line

        B3 = B2 & $0F                   ' Get 4 low order bits
        POKE PORTA, B3                  ' Send B3 to LCD
        PULSOUT 3, 100                  ' Toggle Enable line
        PAUSE 2                         ' Wait 2ms to complete
        POKE PORTA, $10                 ' Set RS = 1
        RETURN

        END                             ' End of program
```

Figure 5.93   PicBasic program listing of Project 20

Project 21

Project title: LCD-based clock with hours–minutes–seconds display

Project description: In this project an LCD-based digital clock is designed. Hours, minutes, and seconds are displayed on the LCD in the following format:

HH:MM:SS

Two push-button switches are used to set the hours and minutes. Pressing the hours button increments the hours between 00 and 23. Similarly, pressing the minutes button increments the minutes between 00 and 59 so that the time can be set.

Figure 5.94 shows the block diagram of the project.

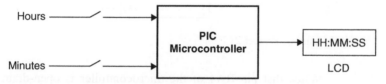

Figure 5.94   Block diagram of Project 21

Hardware: The circuit diagram of the project is shown in Figure 5.95.

A PIC16F627 microcontroller is used in the project with a 4 MHz internal clock and its master clear circuit is enabled during programming of the chip. The LCD is connected in the default mode as described in project 20. Hours and minutes buttons are connected to RB0 and RB1 pins of PORTB, respectively.

The I/O connections are summarised below:

PORT pin	Mode	Connection
RA0	Output	LCD D4
RA1	Output	LCD D5
RA2	Output	LCD D6
RA3	Output	LCD D7
RB3	Output	LCD E
RA4	Output	LCD RS
RB0	Input	Hours button
RB1	Input	Minutes button

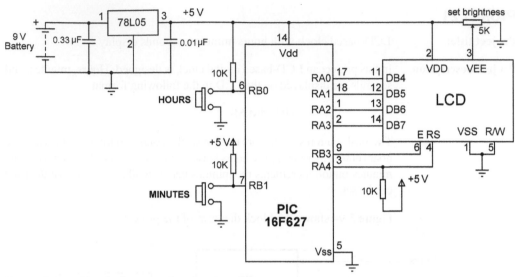

Figure 5.95   Circuit diagram of Project 21

Notice that pin RA4 of the microcontroller is open-drain output and should be connected to the +V supply with a 10K resistor.

Flow diagram: The flow diagram of the project is given in Figure 5.96. The operation of the project is based on a timer interrupt. The timer interrupt is set to generate an interrupt every second. Inside the ISR the time is advanced by 1 s and the minutes and hours are adjusted if necessary and the hours, minutes, and seconds are displayed every second on the LCD.

In addition to displaying the time, the hours and minutes buttons can be used to set the time at the beginning of a session. Pressing the hours button advances the hours display by 1 h. Similarly, pressing the minutes button advances the minutes display by 1 min.

Software: **PicBasic**
The program in this project is based on the timer interrupt. PicBasic language does not support interrupts from high-level language and therefore only the PicBasic Pro program of this project is given.

PicBasic Pro
Figure 5.97 shows the PicBasic Pro program listing of the project. At the beginning of the program *Hrs_button* and *Mins_button* are assigned to RB0 and RB1, respectively. The following variables are then declared:

Hour	Stores the hours field of time
Minute	Stores the minutes field of time

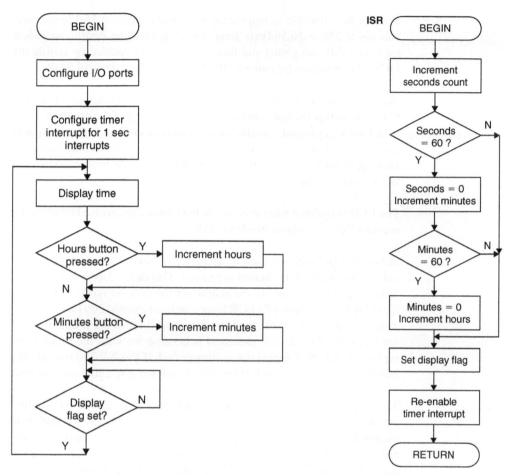

Figure 5.96 Flow diagram of Project 21

Second	Stores the seconds field of time
Ticks	This variable stores the tick number. It is incremented by one whenever a timer interrupt occurs. A second consists of 61 Ticks
Disp	This variable controls writing to the LCD. When Disp = 1, the LCD is updated
Delay	This variable is used in the delay loop of the contact debouncing subroutine

Initially the Hour, Minute, Second, and Ticks are all cleared to zero. Timer pre-scaler is set to 64 by loading the OPTION_REG to bit pattern "00000101" (hexadecimal $05). Timer interrupt register TMR0 is then left to count from 0 to 255 so that the timer interrupts occur at 16.384 ms intervals ($64 \times 256 = 16.384$ ms). Interrupt service routine starting address is named as ISR and global and timer interrupts are enabled by setting the INTCON register to bit pattern "10100000" (i.e. hexadecimal $A0).

Inside the main program the hours and minutes buttons are checked continuously so that the time can be set at the beginning of the program. If the hour button is pressed, variable *Hour* is incremented by 1. When Hour is equal to 24 it is cleared to zero. Similarly, when the minute button is pressed, variable *Minute* is incremented by 1. When Minute reached to 60 it is cleared to zero.

The LCD is updated whenever any of the buttons are pressed or when the seconds field is updated inside the ISR.

Subroutine Debounce is used to debounce the switch contacts. A 200 ms delay is inserted after a button is pressed. This delay debounces the contacts and also gives time to the user to set the time correctly. Notice that the delay loop consists of a FOR loop which is repeated 200 times and the actual delay is 1 ms inside the loop. The reason for doing it this way and not using the PAUSE 200 statement is because we want the timer interrupts to be accepted during the waiting period. If PAUSE 200 is used then interrupts will not be checked for 200 ms and we may get wrong counts.

The interrupt service routine starts with label ISR. Inside this routine variable Ticks is incremented by 1. When Ticks reaches 61 then it is assumed that 1 s has elapsed (61×16.384 ms = 999.424 ms) and variable Second is incremented by 1. When Second reaches 60 it is cleared to zero and Minute is incremented by 1. Similarly, when Minute reaches 60 it is cleared to zero and Hour is incremented by 1. At the end of the ISR variable Disp is set to 1 if the time has been updated and timer interrupts are re-enabled by clearing bit 2 of register INTCON.

Notice that the actual timer interrupt interval is 999.424 ms which is 576 ms short of a second. If we take into account the delay caused by the operations inside the ISR our timer intervals are probably very close to 1 s (it is not possible to calculate the exact delay when using a high-level language since the exact execution times of the instructions are not known).

In this project the internal clock of the microcontroller is used as the clock source. This clock is not accurate and for more accurate results, use of an external crystal clock source is recommended.

```
'**********************************************************************
'
'                    LCD BASED CLOCK
'                    ==================
'
'
' In this project an LCD display is connected to a PIC16F627 microcontroller.
' The microcontroller is configured to operate with a 4MHz internal clock.
'
' The project is a clock, showing the hours, minutes, and seconds in the
' following format:
'
'        HH:MM:SS
'
' The connection between the LCD display and the microcontroller is as follows:
'
'        Display          Microcontroller pin
'        DB4              RA0
'        DB5              RA1
'        DB6              RA2
'        RB7              RA3
'        E                RB3
'        RS               RA4
'
'        Hrs button       RB0
'        Mins button      RB1
'
' Two push-button switches are connected to RB0 and RB1 pins of PORTB
' in order to set the time (hours and minutes fileds of the time).
'
' The hours button is connected to RB0 and pressing this button increments
' hours by 1. When hours reaches 24, it is reset back to 0. Similarly, the
' minutes button is connected to RB1 and pressing this button increments
' the minutes by 1. When minutes reached 60, it is reset back to 0.
'
' The timer interrupt TMR0 is used to update the time. The timer is configured
' to interrupt at every 16.384ms. When the count is 61, one second is elapsed
' and the seconds variable is incremented by 1. The minutes or the hours
' variables are incremented if necessary.
'
' A 10K resistor is used to pull-up pin RA4 of the microcontroller.
'
' RW pin of the LCD is connected to ground. The brightness of the LCD is
' controlled by connecting a 5K variable resistor to pin VEE of the display.
```

Figure 5.97   (Continued)

```
'
'
' Author:            Dogan Ibrahim
' Date:              November, 2005
' Compiler:          PicBasic Pro
' File:              LCD3.BAS
'
' Modifications
' ==========
'
'*********************************************************************
'
' DEFINITIONS
'

Symbol Hrs_button = PORTB.0          ' Hour setting button
Symbol Mins_button = PORTB.1         ' Minute setting button

Ticks   VAR   byte                   ' Tick count (61 ticks = 1 sec)
Hour    VAR   byte                   ' Hour variable
Minute  VAR   byte                   ' Minute variable
Second  VAR   byte                   ' Second variable
Disp    VAR   byte                   ' Disp = 1 to update display
Delay   VAR   byte                   ' Used to Debounce button

        TRISA = 0                    ' PORTA is output
        TRISB = 3                    ' RB0,RB1 are inputs
        CMCON = 7                    ' PORTA digital I/O

        PAUSE 500                    ' Wait 0.5sec for LCD to initialize
'
' Clear Hour, Minute, Second and Ticks to zero
'
        Hour = 0
        Minute = 0
        Second = 0
        Ticks = 0


'
' Initialize timer interrupt. The prescaler is set to 64 and the
' TMR0 is left to run from 0 to 255. With a clock frequency of 4MHz,
' The timer interrupt is generated at every 256 * 64 = 16.384ms.
```

Figure 5.97   (Continued)

' Inside the ISR, variable ticks is incremented by 1. When Ticks = 61
' then time for a timer interrupt is: 61*16.384 = 999.424ms and variable
' Second is then updated. i.e. Second is updated nearly every second.
'

```
        OPTION_REG = $05            ' Set prescaler = 64
        ON INTERRUPT GOTO ISR       ' ISR routine
        INTCON = $A0                ' Enable TMR0 interrupt and global interrupts

        LCDOUT $FE, 1               ' Clear LCD
'
```

' Beginning of MAIN program loop
'

```
LOOP:
'
```

' Check Hour button and if pressed increment variable Hour
'

```
        IF Hrs_button = 0 THEN
                Hour = Hour + 1
                IF Hour = 24 THEN Hour = 0
                Gosub Debounce
        ENDIF
'
```

' Check Minute button and if pressed increment variable Minute
'

```
        IF Mins_button = 0 THEN
                Minute = Minute +1
                IF Minute = 60 THEN Minute = 0
                Gosub Debounce
        ENDIF
'
```

' Display update section. The display is updated when variable
' Disp is 1. This variable is set to 1 inside the ISR when the
' seconds changes. The cursor is set to home position and the
' time is displayed on the LCD
'

```
        IF Disp = 1 THEN
                LCDOUT $FE, 2
                LCDOUT DEC2 Hour, ":",DEC2 Minute, ":",DEC2 Second
                Disp = 0
        ENDIF
        GOTO LOOP
```

Figure 5.97   (Continued)

'

' This subroutine Debounces the buttons. Also, a delay is introduced when
' a button is pressed so that the variable attached to the button (Hour or Second)
' can be incremented after a small delay.
'

```
Debounce:
        FOR Delay = 1 To 200
                Pause 1                ' Delay 1ms inside a loop. This way,
        NEXT Delay                     ' timer interrupts are not stopped
        Disp = 1                       ' Set display flag to 1
        RETURN
```

'

' This is the Timer interrupt Service Routine. The program jumps to this code
' whenever the timer overflows from 255 to 0. i.e. every 256 count. The prescaler
' is set to 64 and the clock frequency is 4MHz. i.e. the basic instruction cycle
' time is 1 microsecond. Thus, timer interrupts occur at every 64*256 = 16.384ms.
' Variable Ticks is incremented by 1 each time a timer interrupt occurs. When Ticks
' is equal to 61, then one second has elapsed (16.384*61 = 999.424ms) and then
' variable Second is incremented by 1. When Second is 60, variable Minute is
' incremented by 1. When Minute is 60, variable Hour is incremented by 1.
'

' Timer TMR0 interrupts are re-enabled just before the program exits this routine.
'

```
DISABLE
ISR:
        Ticks = Ticks + 1
        IF Ticks < 61 THEN NoUpdate
```

'

' 1 second has elapsed, now update seconds and if necessary minutes and hours.

Figure 5.97   (Continued)

'
```
            Ticks = 0
            Second = Second + 1              ' Update second
            IF Second = 60 THEN
                    Second = 0
                    Minute = Minute + 1       ' Update Minute
                    IF Minute = 60 THEN
                            Minute = 0
                            Hour = Hour + 1    ' Update Hour
                            IF Hour = 24 THEN
                                    Hour = 0
                            ENDIF
                    ENDIF
            ENDIF

            Disp = 1                          ' Set to update display
'
' End of time update
'
NoUpdate:
            INTCON.2 = 0                      ' Re-enable TMR0 interrupts
            Resume
            ENABLE                            ' Re-enable interrupts

            END

            END                               ' End of program
```

Figure 5.97   PicBasic Pro program of Project 21

Project 22

Project title: LCD-based chronometer

Project description: In this project an LCD-based chronometer is designed. The chronometer counts the elapsed time in seconds and displays in hours, minutes, and seconds in the following format:

HH:MM:SS

Three push-button switches are used to start, stop, and clear the chronometer. Pressing button START starts the chronometer which counts in seconds. Pressing button STOP stops the counting. Pressing button CLEAR clears the display so that the chronometer is ready for the next count.

Figure 5.98 shows the block diagram of the project.

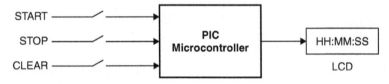

Figure 5.98   Block diagram of Project 22

Hardware: The circuit diagram of the project is shown in Figure 5.99.

A PIC16F627 microcontroller is used in the project with a 4 MHz internal clock. The LCD is connected in the default mode as described in project 21. START, STOP, and CLEAR buttons are connected to RB0, RB1, and RB2 pins of PORTB, respectively.

The I/O connections are summarised below:

PORT pin	Mode	Connection
RA0	Output	LCD D4
RA1	Output	LCD D5
RA2	Output	LCD D6
RA3	Output	LCD D7
RB3	Output	LCD E
RA4	Output	LCD RS
RB0	Input	START button
RB1	Input	STOP button
RB2	Input	CLEAR button

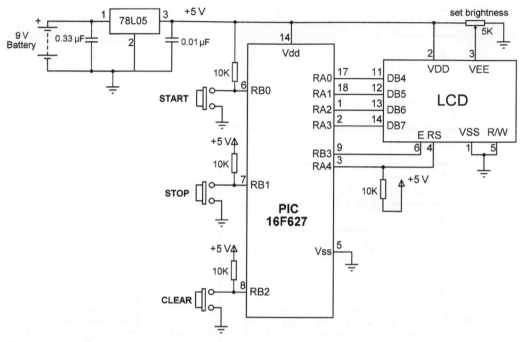

Figure 5.99   Circuit diagram of Project 22

Notice that pin RA4 of the microcontroller is open-drain output and should be connected to the +V supply with a 10K resistor.

Flow diagram: The flow diagram of the project is given in Figure 5.100. The operation of the project is based on a timer interrupt. The timer interrupt is set to generate an interrupt every second when the chronometer is started. Pressing the START button clears the timer register TMR0 and enables interrupts. Pressing STOP button disables interrupts so that the final count can be displayed and viewed on the LCD. Pressing the CLEAR button clears the hours, minutes, seconds, and ticks so that a new count can start.

Software: **PicBasic**
The program in this project is based on the timer interrupt. PicBasic language does not support interrupts from high-level language and therefore only the PicBasic Pro program of this project is given.

PicBasic Pro
Figure 5.101 shows the PicBasic Pro program listing of the project. At the beginning of the program *START_button, STOP_button*, and *CLEAR_button*

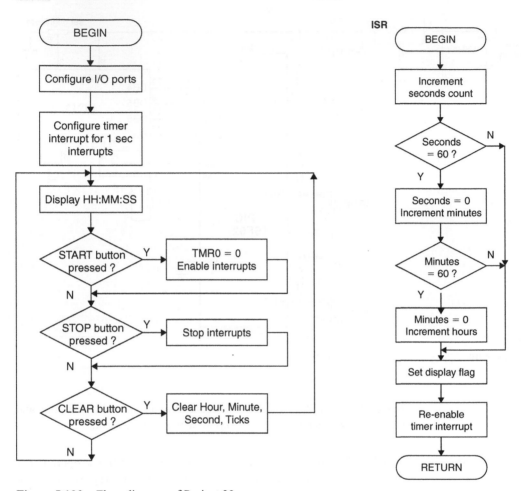

Figure 5.100   Flow diagram of Project 22

are assigned to RB0, RB1, and RB2, respectively. The following variables are then declared:

Hour   Stores the hours field of time
Minute   Stores the minutes field of time
Second Stores the seconds field of time
Ticks   This variable stores the tick number. It is incremented by one whenever a timer interrupt occurs. A second consists of 61 Ticks
Disp   This variable controls writing to the LCD When Disp = 1, the LCD is updated

Initially the Hour, Minute, Second, and Ticks are all cleared to zero. Timer pre-scaler is set to 64 by loading the OPTION_REG to bit pattern "00000101" (hexadecimal $05). Timer interrupt register TMR0 is then left to count from 0 to 255 so that the timer interrupts occur at 16.384 ms intervals ($64 \times 256 = 16.384$ ms).

When the START button is pressed, timer register TMR0 is reset to zero and timer interrupts are enabled. Thus, an interrupt is generated every second and the display shows the elapsed time in HH:MM:SS. When the STOP button is pressed timer interrupts are disabled and the final count is displayed on the LCD. Pressing the CLEAR button clears the time variables so that the chronometer is ready for the next count.

The interrupt service routine starts with label ISR. Inside this routine variable Ticks is incremented by 1. When Ticks reaches 61 then it is assumed that 1 s has elapsed (61×16.384 ms = 999.424 ms) and variable Second is incremented by 1. When Second reaches 60 it is cleared to zero and Minute is incremented by 1. Similarly, when Minute reaches 60 it is cleared to zero and Hour is incremented by 1. At the end of the ISR variable Disp is set to 1 if the time has been updated and timer interrupts are re-enabled by clearing bit 2 of register INTCON.

Notice that the actual timer interrupt interval is 999.424 ms which is 576 ms short of a second. If we take into account the delay caused by the operations inside the ISR our timer intervals are probably very close to 1 s (it is not possible to calculate the exact delay when using a high-level language since the exact execution times of the instructions are not known).

In this project the internal clock of the microcontroller is used as the clock source. The internal clock is not accurate and for more accurate results, use of an external crystal clock source is recommended.

```
'**********************************************************************
'
'                   LCD BASED CHRONOMETER
'                   ==========================
'
' In this project an LCD display is connected to a PIC16F627 microcontroller.
' The microcontroller is configured to operate with a 4MHz internal clock. For
' more accurate results, an external crystal clock source should be used.
'
' The project is a chronometer, counting in seconds and displaying the hours,
' minutes, and seconds in the following format:
'
'       HH:MM:SS
```

Figure 5.101   (Continued)

```
'
' The connection between the LCD display and the microcontroller is as follows:
'
'         Display          Microcontroller pin
'         DB4              RA0
'         DB5              RA1
'         DB6              RA2
'         RB7              RA3
'         E                RB3
'         RS               RA4
'
'         START button     RB0
'         STOP button      RB1
'         CLEAR button     RB2
'
' Three push-button switches are connected to RB0, RB1 and RB2 pins of
' PORTB. Pressing START starts the chronometer counting in seconds. Pressing
' STOP button stops the chronometer and displays the elapsed time in HH:MM:SS
' format. Pressing the CLEAR button clears the chronometer so that it is ready for
' the next count.
'
' The timer interrupt TMR0 is used to update the time. The timer is configured
' to interrupt at every 16.384ms. When the count is 61, one second is elapsed
' and the seconds variable is incremented by 1. The minutes or the hours
' variables are incremented if necessary.
'
' A 10K resistor is used to pull-up pin RA4 of the microcontroller.
'
' RW pin of the LCD is connected to ground. The brightness of the LCD is
' controlled by connecting a 5K variable resistor to pin VEE of the display.
'
'
' Author:        Dogan Ibrahim
' Date:          November, 2005
' Compiler:      PicBasic Pro
' File:          LCD4.BAS
'
' Modifications
' ===========
'
'*********************************************************************************
```

Figure 5.101   (Continued)

```
'
' DEFINITIONS
'
Symbol START_button = PORTB.0        ' START button
Symbol STOP_button = PORTB.1         ' STOP button
Symbol CLEAR_button = PORTB.2        ' CLEAR button

Ticks    VAR    byte                 ' Tick count (61 ticks = 1 sec)
Hour     VAR    byte                 ' Hour variable
Minute   VAR    byte                 ' Minute variable
Second   VAR    byte                 ' Second variable
Disp     VAR    byte                 ' Disp = 1 to update display
Delay    VAR    byte                 ' Used to Debounce button

         TRISA = 0                   ' PORTA is output
         TRISB = 7                   ' RB0,RB1,RB2 are inputs
         CMCON = 7                   ' PORTA digital I/O

         PAUSE 500                   ' Wait 0.5 sec for LCD to initialize
'
' Clear Hour, Minute, Second and Ticks to zero
'
         Hour = 0                    ' Clear hours
         Minute = 0                  ' Clear minutes
         Second = 0                  ' Clear seconds
         Ticks = 0                   ' Clear ticks
         Disp = 1                    ' Force to display 00:00:00 at startup

'
' Initialize timer interrupt. The prescaler is set to 64 and the
' TMR0 is left to run from 0 to 255. With a clock frequency of 4MHz,
' The timer interrupt is generated at every 256 * 64 = 16.384ms.
' Inside the ISR, variable ticks is incremented by 1. When Ticks = 61
' then time for a timer interrupt is: 61*16.384 = 999.424ms and variable
' Second is then updated. i.e. Second is updated nearly every second.
'
         OPTION_REG = $05            ' Set prescaler = 64
         ON INTERRUPT GOTO ISR       ' ISR routine
         LCDOUT $FE, 1               ' Clear LCD
'
' Beginning of MAIN program loop
```

Figure 5.101 (Continued)

```
'
LOOP:
'
' Check if START button is pressed and enable timer interrupts so that
' counting starts if this button is pressed
'
        IF START_button = 0 THEN
                TMR0 = 0                 ' Initialize TMR0 register
                INTCON = $A0             ' Enable timer interrupt
                Disp = 1                 ' Enable display
        ENDIF
'
' Check if STOP button is pressed and disable timer interrupt so that
' counting stops and displays the elapsed time in HH:MM:SS format
'
        IF STOP_button = 0 THEN
                INTCON = 0               ' Disable timer interrupt
                Disp = 1                 ' Enable display
        ENDIF
'
' Check if CLEAR button is pressed and clear the display and time variables
' so that a new count can start.
'
        IF CLEAR_button = 0 THEN
                Hour = 0
                Minute = 0
                Second = 0
                Ticks = 0
                Disp = 1
        ENDIF
'
' Display update section. The display is updated when variable Disp = 1.
' This variable is set to 1 inside the ISR when the seconds changes.
' The cursor is set to home position and the time is displayed on the LCD.
'
        IF Disp = 1 THEN
                LCDOUT $FE, 2
                LCDOUT DEC2 Hour, ":",DEC2 Minute, ":",DEC2 Second
                Disp = 0
        ENDIF
        GOTO LOOP


'
' This is the Timer interrupt Service Routine. The program jumps to this code
```

Figure 5.101   (Continued)

' whenever the timer overflows from 255 to 0. i.e. every 256 count. The prescaler
' is set to 64 and the clock frequency is 4MHz. i.e. the basic instruction cycle
' time is 1 microsecond. Thus, timer interrupts occur at every 64*256 = 16.384ms.
' Variable Ticks is incremented by 1 each time a timer interrupt occurs. When Ticks
' is equal to 61, then one second has elapsed (16.384*61 = 999.424ms) and then
' variable Second is incremented by 1. When Second is 60, variable Minute is
' incremented by 1. When Minute is 60, variable Hour is incremented by 1.
'

' Timer TMR0 interrupts are re-enabled just before the program exits this routine.
'

```
DISABLE
ISR:
        Ticks = Ticks + 1
        IF Ticks < 61 THEN NoUpdate
'
' 1 second has elapsed, now update seconds and if necessary minutes and hours.
'
        Ticks = 0
        Second = Second + 1
        IF Second = 60 THEN
                Second = 0
                Minute = Minute + 1
                IF Minute = 60 THEN
                        Minute = 0
                        Hour = Hour + 1
                        IF Hour = 24 THEN
                                Hour = 0
                        ENDIF
                ENDIF
        ENDIF

        Disp = 1                                ' Set to update display
'
' End of time update
'
NoUpdate:
        INTCON.2 = 0                            ' Re-enable TMR0 interrupts
        Resume
        ENABLE                                  ' Re-enable interrupts

        END

        END                                     ' End of program
```

Figure 5.101   PicBasic Pro program of Project 22

Project 23

Project title: LCD-based voltmeter using A/D converter

Project description: In this project an LCD-based voltmeter is designed. The project can be used to measure and display analog voltages to up to +5 V. The voltage is displayed in millivolts in the following format:

$$V = nnnn$$

where nnnn is the measured voltage. Figure 5.102 shows the block diagram of the project where the voltage to be measured is applied to one of the analog-to-digital converter (A/D) channels of a PIC microcontroller having built-in A/D converters.

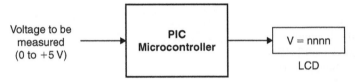

Figure 5.102   Block diagram of Project 23

Hardware: The circuit diagram of the project is shown in Figure 5.103. In this project a PIC16F73-type microcontroller is used. This is a 28-pin microcontroller with built-in 5 channel A/D converters, each having 8-bits of resolution. Other PIC microcontrollers such as PIC16F630 or PIC16F877, or others with built-in A/D converters can easily be used in this project.

PIC16F73 is a 28-pin microcontroller with the following features:

- 8 K flash program memory
- 368 bytes RAM memory
- Up to 20 MHz operation
- 3 timer circuits
- Analog capture, compare and PWM circuits
- 8-bit 5 channel A/D converter
- Built-in USART
- SPI and I²C bus compatibility.

In this project, the microcontroller is operated from a 4 MHz resonator and the voltage to be measured is applied to analog input AN0 of the microcontroller. The analog channels are named AN0 to AN4 and they correspond to the following PORTA names:

Pin	Channel
RA0	AN0
RA1	AN1
RA2	AN2
RA3	AN3
RA4	AN4

The default LCD connections also use pins RA0 to RA4. In order to reserve pins RA0 to RA4 for analog channels, the LCD is connected to PORTB as shown below.

PORTB	LCD pin
RB0	D4
RB1	D5
RB2	D6
RB3	D7
RB4	E
RB5	RS

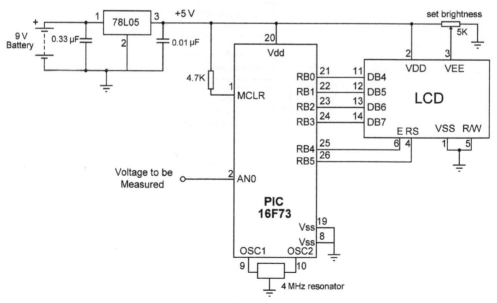

Figure 5.103 Circuit diagram of Project 23

The new LCD connection is defined using a set of DEFINE statements as described in the programming section. The operation of the project is simple: Analog voltage is sampled every second and converted into digital form. The voltage is then scaled and displayed on the LCD.

Flow diagram: The flow diagram of the project is given in Figure 5.104. At the beginning of the program, LCD connections, port directions, and the A/D converter are configured. The voltage to be measured is then converted into digital form, scaled and displayed on the LCD. After 1 s delay this process is repeated.

Software: **PicBasic**
The PicBasic program of this project is complex since LCDs are not supported directly and the LCD routines developed in Project 20 use the default LCD connections. Only the PicBasic Pro program listing of this project is given.

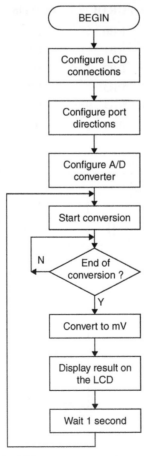

Figure 5.104   Flow diagram of Project 23

PicBasic Pro

The PicBasic Pro program listing of the project is given in Figure 5.105. At the beginning of the program a set of DEFINE statements are used to define the connections between the LCD and the microcontroller.

Variable *Res* stores the converted digital data. Variable *Volts* stores the result of conversion in millivolts. In order to convert the measured voltage to millivolts, it is necessary to multiply the result of the conversion with 19.53 (256 steps correspond to 5000 mV, thus, each step is 5000 mV/256 = 19.53 mV). But since the PicBasic Pro language does not support floating point arithmetic, an approximation is made here and the result is multiplied with 19 only.

The steps for an A/D conversion using the PIC16F73 microcontroller are as follows (assuming that the A/D conversion interrupt is not used):

- Configure the A/D module
  - Configure analog pins, reference voltage, and digital I/O (register ADCON1)
  - Select A/D conversion clock (register ADCON0)
  - Turn on A/D module (register ADCON0)
- Select an A/D input channel
- Start A/D conversion
  - Set GO/DONE bit of register ADCON0
- Wait for the conversion to complete
  - Wait until GO/DONE bit of register ADCON0 is cleared
- Read the A/D result register (ADRES)
- Go to step 2 or 3 for next conversion.

In Figure 5.105, register ADCON1 is initially cleared so that AN0 to AN4 are analog inputs and the A/D converter reference is the supply voltage, VDD. Register ADCON0 is then set to bit pattern "11000001" to select the internal RC oscillator as the source of clock for the A/D. Analog channel AN0 is also selected and the A/D module is turned on.

A/D conversion is started by setting the GO/DONE bit of register ADCON0, i.e. ADCON0.2 = 1. The program then waits until the conversion is complete which is indicated by the GO/DONE bit going to logic 0, i.e. ADCON0.2 = 0. The result of the conversion is then read from register ADRES and is stored as a digital value between 0 and 255 in variable *Res*. *Res* is multiplied with 19 (it should be 19.53 for an exact result) and stored in variable *Volts*. *Volts* stores the measured voltage in millivolts. This voltage is then displayed on the LCD as a 4-digit decimal number.

The program repeats after a one-second delay.

```
'***********************************************************************
'
'                    LCD BASED VOLTMETER
'                    =======================
'
' In this project an LCD display is connected to a PIC16F73 microcontroller.
' The microcontroller is configured to operate with a 4MHz external resonator.
'
' The project is a voltmeter, which can measure the voltage applied to the analog
' input AN0. The voltage to be measured must be between 0 V and +5V.
'
' The connection between the LCD display and the microcontroller is as follows:
'
'            Display          Microcontroller pin
'            DB4              RB0
'            DB5              RB1
'            DB6              RB2
'            RB7              RB3
'            E                RB4
'            RS               RB5
'
'            Analog input     AN0 (RA0)
'
' RW pin of the LCD is connected to ground. The brightness of the LCD is
' controlled by connecting a 5K variable resistor to pin VEE of the display.
'
' The PIC16F73 microcontroller has built in 8-bit 5 channel A/D converters.
' The A/D reference voltage is set to +5V. With 8-bit converters, operating with
' a reference voltage of +5V, the bit resolution is 5000/256 = 19.53mV.
'
' The result is displayed in mV in the following format:
'
'         V = nnnn
'
' Author:      Dogan Ibrahim
' Date:        November, 2005
' Compiler:    PicBasic Pro
' File:        LCD5.BAS
'
' Modifications
' ===========
'
'***********************************************************************
```

Figure 5.105   (Continued)

```
'
' DEFINITIONS
'
' Define LCD connections
'
DEFINE LCD_DREG   PORTB               ' LCD Data bits on PORTB
DEFINE LCD_DBIT          0            ' PORTB starting address
DEFINE LCD_RSREG PORTB                ' LCD RS bit on PORTB
DEFINE LCD_RSBIT         5            ' LCD RS bit address
DEFINE LCD_EREG   PORTB               ' LCD E bit on PORTB
DEFINE LCD_EBIT          4            ' LCD E bit address
DEFINE LCD_BITS          4            ' LCD in 4-bit mode
DEFINE LCD_LINES         2            ' LCD has 2 rows


Res     Var Word                     ' A/D converter result
Volts   Var Word                     ' Result of conversion in mV
Conv    Con 19                       ' 5000/256 = 19.53, take 19

        TRISA = 1                    ' RA0 (AN0) is input
        TRISB = 0                    ' PORTB is output

        PAUSE 500                    ' Wait 0.5sec for LCD to initialize
'
' Initialize the A/D converter
'
        ADCON1 = 0                   ' Make AN0 to AN4 as analog inputs,
                                     ' make reference voltage = VDD
        ADCON0 = %11000001           ' A/D clock is internal RC, select AN0
                                     ' Turn on A/D converter
        LCDOUT $FE, 1                ' Clear LCD

AGAIN:
'
' Start A/D conversion
'
        ADCON0.2 = 1
'
' Wait until conversion is complete
'
WT:     PAUSE 1
        IF ADCON0.2 = 1 THEN WT
        Res = ADRES                  ' Get result of conversion
        Volts = Res * Conv           ' Result in mV
```

Figure 5.105   (Continued)

```
LCDOUT $FE,2,"V = ",DEC4 Volts        ' Display result
PAUSE 1000                            ' Wait 1 second
GOTO AGAIN                            ' Repeat

END                                  ' End of program
```

Figure 5.105   PicBasic Pro listing of Project 23

The displayed voltage by the program in Figure 5.105 is not accurate since the converted signal is multiplied by 19 and not by 19.53. The reason for this was because the PicBasic Pro language does not support floating point arithmetic. The result could however be made more accurate by performing the multiplication with 19.53 as follows:

- Consider 19.53 as 19 + 0.53
- Read the A/D result into variable *Res*
- Multiply *Res* with 19 and store in *Volts1*
- Multiply *Res* with 53 and store in *Volts2*
- Divide *Volts2* to *Volts*100
- Add *Volts2* to *Volts1*. *Volts1* now contains a number which is more closely related to *Res*19.53*

The program given in Figure 5.106 implements the changes described above. In this program variable Volts1 stores the measured voltage in millivolts and this variable is displayed as a 4 digit decimal number.

```
'*********************************************************************
'
'                    LCD BASED VOLTMETER
'                    =======================
'
'
' In this project an LCD display is connected to a PIC16F73 microcontroller.
' The microcontroller is configured to operate with a 4MHz external resonator.
'
' The project is a voltmeter, which can measure the voltage applied to the analog
' input AN0. The voltage to be measured must be between 0 V and +5 V.
'
' The connection between the LCD display and the microcontroller is as follows:
'
'       Display         Microcontroller pin
'       DB4             RB0
```

Figure 5.106   (Continued)

```
'      DB5          RB1
'      DB6          RB2
'      RB7          RB3
'      E            RB4
'      RS           RB5
'
'      Analog input    AN0 (RA0)
'
' RW pin of the LCD is connected to ground. The brightness of the LCD is
' controlled by connecting a 5K variable resistor to pin VEE of the display.
'
' The PIC16F73 microcontrolelr has built in 8-bit 5 channel A/D converters.
' The A/D reference voltage is set to +5 V. With 8-bit converters, operating with
' a reference voltage of +5 V, the bit resolution is 5000/256 = 19.53 mV.
'
' The result is displayed in mV in the following format:
'
'      V = nnnn
'
' This program is similar to LCD5.BAS, but here the result is more accurate since
' the conversion factor is taken as 19.53 and not just 19.
'
'
' Author:       Dogan Ibrahim
' Date:         November, 2005
' Compiler:     PicBasic Pro
' File:         LCD5-1.BAS
'
' Modifications
' ===========
'
'**********************************************************************
'
' DEFINITIONS
'
' Define LCD connections
'
DEFINE LCD_DREG    PORTB          ' LCD Data bits on PORTB
DEFINE LCD_DBIT           0       ' PORTB starting address
DEFINE LCD_RSREG   PORTB          ' LCD RS bit on PORTB
DEFINE LCD_RSBIT          5       ' LCD RS bit address
DEFINE LCD_EREG    PORTB          ' LCD E bit on PORTB
DEFINE LCD_EBIT           4       ' LCD E bit address
```

Figure 5.106   (Continued)

```
DEFINE LCD_BITS        4              ' LCD in 4-bit mode
DEFINE LCD_LINES       2              ' LCD has 2 rows

Res     Var Word                      ' A/D converter result
Volts1  Var Word                      ' First part of result in mV
Volts2  Var Word                      ' Second part of result in mV

Conv1   Con 19                        ' 5000/256 = 19.53, this is the decimal part
Conv2   Con 53                        ' This is the fractional part

        TRISA = 1                     ' RA0 (AN0) is input
        TRISB = 0                     ' PORTB is output

        PAUSE 500                     ' Wait 0.5sec for LCD to initialize
'
' Initialize the A/D converter
'
        ADCON1 = 0                    ' Make AN0 to AN4 as analog inputs,
                                      ' make reference voltage = VDD
        ADCON0 = %11000001           ' A/D clock is internal RC, select AN0
                                      ' Turn on A/D converter
        LCDOUT $FE, 1                 ' Clear LCD

AGAIN:
'
' Start A/D conversion
'
        ADCON0.2 = 1
'
' Wait until conversion is complete
'
WT:     PAUSE 1
        IF ADCON0.2 = 1 THEN WT
        Res = ADRES                   ' Get result of conversion

        Volts1 = Res * Conv1          ' Multiply by 19
        Volts2 = Res * Conv2          ' Multiply by 53
        Volts2 = Volts2 / 100
        Volts1 = Volts1 + Volts2      ' Result in mV
        LCDOUT $FE,2,"V = ",DEC4 Volts1   ' Display result
        PAUSE 1000                    ' Wait 1 second
        GOTO AGAIN                    ' Repeat

        END                           ' End of program
```

Figure 5.106   More accurate PicBasic Pro program

PicBasic Pro language provides a high-level instruction called ADCIN for starting an A/D conversion and reading the result of the conversion. The format of this instruction is

ADCIN channel, var

Where *channel* is the A/D channel used, and *var* is the variable which is to store the result of the conversion. Using this instruction, simplifies the programming of an A/D converter channel. Figure 5.107 gives the program listing which makes use of the ADCIN instruction.

Notice that the width of the A/D is defined with ADC_BITS, the A/D clock is defined with ADC_CLOCK (3 corresponds to the internal RC clock), and the A/D sampling time is defined with ADC_SAMPLEUS. Although we are using the ADCIN statement to read the analog input, we still have to configure the ADCON0 and ADCON1 registers before starting a conversion.

```
'****************************************************************
'
'                 LCD BASED VOLTMETER
'                 =========================
'
' In this project an LCD display is connected to a PIC16F73 microcontroller.
' The microcontroller is configured to operate with a 4MHz external resonator.
'
' The project is a voltmeter, which can measure the voltage applied to the analog
' input AN0. The voltage to be measured must be between 0V and +5V.
'
' The connection between the LCD display and the microcontroller is as follows:
'
'       Display         Microcontroller pin
'       DB4             RB0
'       DB5             RB1
'       DB6             RB2
'       RB7             RB3
'       E               RB4
'       RS              RB5
'
'       Analog input    AN0 (RA0)
'
' RW pin of the LCD is connected to ground. The brightness of the LCD is
' controlled by connecting a 5K variable resistor to pin VEE of the display.
```

Figure 5.107   (Continued)

```
'
' The PIC16F73 microcontroller has built in 8-bit 5 channel A/D converters.
' The A/D reference voltage is set to +5V. With 8-bit converters, operating with
' a reference voltage of +5V, the bit resolution is 5000/256 = 19.53mV.
'
' The result is displayed in mV in the following format:
'
'          V = nnnn
'
' This program is similar to LCD5.BAS, but here the result is more accurate since
' the conversion factor is taken as 19.53 and not just 19.
'
'
' In this program PicBasic statement ADCIN is used to read analog data
'
'
' Author:        Dogan Ibrahim
' Date:          November, 2005
' Compiler:      PicBasic Pro
' File:          LCD5-2.BAS
'
' Modifications
' ===========
'
'*********************************************************************************
'
' DEFINITIONS
'
' Define LCD connections
'
DEFINE LCD_DREG    PORTB                    ' LCD Data bits on PORTB
DEFINE LCD_DBIT          0                  ' PORTB starting address
DEFINE LCD_RSREG   PORTB                    ' LCD RS bit on PORTB
DEFINE LCD_RSBIT         5                  ' LCD RS bit address
DEFINE LCD_EREG    PORTB                    ' LCD E bit on PORTB
DEFINE LCD_EBIT          4                  ' LCD E bit address
DEFINE LCD_BITS          4                  ' LCD in 4-bit mode
DEFINE LCD_LINES         2                  ' LCD has 2 rows
'
' Define A/D converter parameters
'
DEFINE ADC_BITS          8                  ' A/D number of bits
DEFINE ADC_CLOCK         3                  ' Use A/D internal RC clock
DEFINE ADC_SAMPLEUS     50                  ' Set sampling time in us
```

Figure 5.107   (Continued)

```
'
' Variables
'
Res       Var Word                              ' A/D converter result
Volts1    Var Word                              ' First part of result in mV
Volts2    Var Word                              ' Second part of result in mV


'
' Constants
'
Conv1     Con 19                                ' 5000/256 = 19.53, this is the decimal part
Conv2     Con 53                                ' This is the fractional part

          TRISA = 1                             ' RA0 (AN0) is input
          TRISB = 0                             ' PORTB is output

          PAUSE 500                             ' Wait 0.5sec for LCD to initialize
'
' Initialize the A/D converter
'
          ADCON1 = 0                            ' Make AN0 to AN4 as analog inputs,
                                                ' make reference voltage = VDD
          ADCON0 = %11000001                    ' A/D clock is internal RC, select channel AN0
                                                ' Turn on A/D converter
          LCDOUT $FE, 1                         ' Clear LCD

AGAIN:
'
' Start A/D conversion
'
          ADCIN 0, Res                          ' Read Channel 0 data

          Volts1 = Res * Conv1                  ' Multiply by 19
          Volts2 = Res * Conv2                  ' Multiply by 53
          Volts2 = Volts2 / 100
          Volts1 = Volts1 + Volts2              ' Result in mV
          LCDOUT $FE,2,"V = ",DEC4 Volts1       ' Display result
          PAUSE 1000                            ' Wait 1 second
          GOTO AGAIN                            ' Repeat

          END                                   ' End of program
```

Figure 5.107   PicBasic program using the ADCIN instruction

Project 24

Project title: LCD-based thermometer using A/D converter

Project description: In this project an LCD-based thermometer is designed. The project can be
used to display the temperature in degrees centigrade every second in the
following format:

$$\text{Temp} = \text{nnC}$$

Where nn is the measured temperature. Figure 5.108 shows the block diagram of the project where the temperature sensor is connected to one of
the analog-to-digital converter (A/D) channels of a PIC microcontroller.

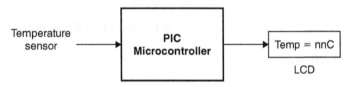

Figure 5.108   Block diagram of Project 24

Hardware: The circuit diagram of the project is shown in Figure 5.109.

In this project a PIC16F73-type microcontroller is used. This is a 28-pin
microcontroller with built-in 5 channel A/D converters, each having 8-
bits of resolution. The microcontroller is operated from a 4 MHz resonator. The temperature sensor used is the LM35DZ (see Figure 5.110)
3-pin analog sensor with a range of 0°C to +100°C. LM35DZ provides
an analog output voltage which is proportional to the measured temperature. The device has 3 pins: Vs, Gnd, and Vo. Vs and Gnd are connected to
the supply voltage and the ground, respectively. It is recommended by the
manufacturers to use a 10 Ω resistor and a 1 μF capacitor filter at the output of the sensor to minimise electrical noise. Vo is the analog output voltage given by

$$V\text{o} = 10\,\text{mV/°C}$$

For example, at a temperature of 20°C the output voltage is 200 mV. In
this project LM35DZ is connected to analog input AN0 of the PIC16F73
microcontroller.

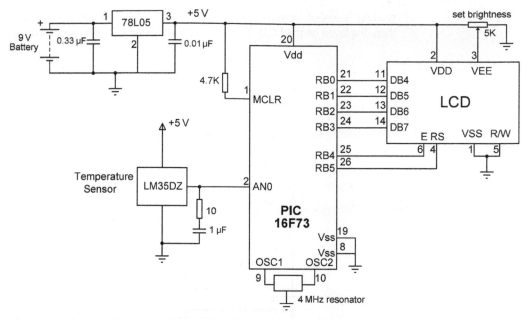

Figure 5.109   Circuit diagram of Project 24

Figure 5.110   LM35DZ temperature sensor

The operation of the project is very simple: the output of the temperature sensor is converted into digital, scaled, and then displayed on the LCD. This process is repeated after one-second delay.

Flow diagram: The flow diagram of the project is given in Figure 5.111. At the beginning of the program LCD connections, port directions and the A/D converter are configured. Analog temperature is then read and converted into digital. The reading is scaled, converted into true degrees centigrade temperature and then displayed on the LCD. This process is repeated after one-second delay.

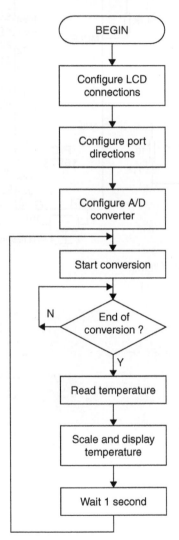

Figure 5.111   Flow diagram of Project 24

Software: **PicBasic**
                       The PicBasic program of this project is complex since LCDs are not supported directly and the LCD routines developed in Project 20 use the

default LCD connections. Only the PicBasic Pro program listing of this project is given.

PicBasic Pro

The PicBasic Pro program listing of the project is given in Figure 5.112. At the beginning of the program LCD connections and the A/D parameters are defined. Variable *Res* stores the converted data. A/D conversion is started using the ADCIN statement. When the conversion is complete, the converted data is available in register *Res*. The contents of *Res* can be converted into millivolts by multiplying it by 19.53 as described in Project 23. But, since the output of the sensor is 10 mV/°C, it will be necessary to divide *Res* by 10 in order to find the real absolute temperature in degrees centigrade. Thus, the temperature can be obtained by the following operation:

$$Res * 19.53/10 = Res * 1.953 \approx 2 * Res$$

In the program, variable *Res* is multiplied by 2 to obtain the temperature with a ±1°C accuracy (the resolution of the A/D converter is 19.53 mV which is equivalent to nearly 2°C). The value of *Res* is then displayed on the LCD as a two-digit decimal number. The above process is repeated after one-second delay.

For more accurate temperature measurements an A/D converter with a higher resolution will be required, e.g. 10-bit or higher.

```
'********************************************************************
'
'                    LCD BASED THERMOMETER
'                    =========================
'
'
' In this project an LCD display is connected to a PIC16F73 microcontroller.
' The microcontroller is configured to operate with a 4MHz external rezonator.
'
' The project is a thermometer, which can measure the environmental
' temperature and then display on the LCD.
'
' A LM35DZ type analog output temperature sensor is used in this project.
' LM35DZ provides an output voltage proportional to the measured temperature.
```

Figure 5.112   (Continued)

```
'
' The connection between the LCD display and the microcontroller is as follows:
'
'        Display      Microcontroller pin
'        DB4          RB0
'        DB5          RB1
'        DB6          RB2
'        RB7          RB3
'        E            RB4
'        RS           RB5
'
'        LM35DZ       AN0 (RA0)
'
' RW pin of the LCD is connected to ground. The brightness of the LCD is
' controlled by connecting a 5K variable resistor to pin VEE of the display.
'
' The PIC16F73 microcontroller has built in 8-bit 5 channel A/D converters.
' The A/D reference voltage is set to +5V. With 8-bit converters, operating with
' a reference voltage of +5V, the bit resolution is 5000/256 = 19.53mV.
'
' The temperature is displayed in degrees C in the following format:
'
'        TEMP = nnC
'
' In this program PicBasic statement ADCIN is used to read analog data
'
'
' Author:      Dogan Ibrahim
' Date:        November, 2005
' Compiler:    PicBasic Pro
' File:        TEMP.BAS
'
' Modifications
' ===========
'
'********************************************************************
'
' DEFINITIONS
'
' Define LCD connections
'
DEFINE LCD_DREG    PORTB                ' LCD Data bits on PORTB
DEFINE LCD_DBIT           0             ' PORTB starting address
```

Figure 5.112   (Continued)

```
DEFINE LCD_RSREG  PORTB          ' LCD RS bit on PORTB
DEFINE LCD_RSBIT       5         ' LCD RS bit address
DEFINE LCD_EREG    PORTB         ' LCD E bit on PORTB
DEFINE LCD_EBIT        4         ' LCD E bit address
DEFINE LCD_BITS        4         ' LCD in 4-bit mode
DEFINE LCD_LINES       2         ' LCD has 2 rows
'
' Define A/D converter parameters
'
DEFINE ADC_BITS        8         ' A/D number of bits
DEFINE ADC_CLOCK       3         ' Use A/D internal RC clock
DEFINE ADC_SAMPLEUS    50        ' Set sampling time in us
'
' Variables
'
Res      Var Word                ' A/D converter result
Temp1    Var Byte                ' Temperature in degrees C

         TRISA = 1               ' RA0 (AN0) is input
         TRISB = 0               ' PORTB is output

         PAUSE 500               ' Wait 0.5sec for LCD to initialize
'
' Initialize the A/D converter
'
         ADCON1 = 0              ' Make AN0 to AN4 as analog inputs,
                                 ' make reference voltage = VDD
         ADCON0 = %11000001      ' A/D clock is internal RC, select AN0
                                 ' Turn on A/D converter
         LCDOUT $FE, 1           ' Clear LCD

AGAIN:
'
' Start A/D conversion
'
         ADCIN 0, Res                             ' Read Channel 0 data
         Temp1 = 2 * Res                          ' Convert to degrees C
         LCDOUT $FE,2,"TEMP = ",DEC2 Temp1, "C"   ' Display decimal part
         PAUSE 1000                               ' Wait 1 second
         GOTO AGAIN                               ' Repeat

         END                                      ' End of program
```

Figure 5.112   PicBasic Pro listing of Project 24

Project 25

Project title: Serial LCD-based thermometer with external EEPROM memory

Project description: In this project an LCD-based thermometer is designed. The project consists of a temperature sensor, a serial LCD display, a PIC microcontroller, and an external I²C bus compatible EEPROM memory. Temperature is measured every minute and stored in the EEPROM memory. During this time the following message is displayed on the LCD:

COLLECTING DATA

After 1 h the measurement stops and the program reads from the EEPROM memory to find the maximum temperature. The maximum temperature is displayed on the LCD in the following format:

Max=nnC

where nn is the measured maximum temperature in 1 h. Figure 5.113 shows the block diagram of the project where the temperature sensor is connected to one of the analog-to-digital converter (A/D) channels of a PIC microcontroller.

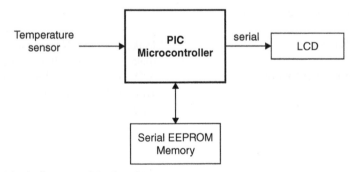

Figure 5.113 Block diagram of Project 25

Hardware: The circuit diagram of the project is shown in Figure 5.114. In this project a PIC16F73-type microcontroller is used. This is a 28-pin microcontroller with built-in 5 channel A/D converters, each having 8-bits of resolution. The microcontroller is operated from a 4 MHz resonator. The temperature sensor used is the LM35DZ 3-pin analog sensor (see Project 24) with a range of 0°C to +100°C. LM35DZ provides an analog output voltage which is proportional to the measured temperature. The device has 3 pins: Vs, Gnd, and Vo. Vs and Gnd are connected to the supply voltage and the ground, respectively. It is recommended by the manufacturers to use a $10\,\Omega$ resistor and a $1\,\mu F$ capacitor filter at the output of the sensor to minimise electrical noise. Vo is the analog output voltage given by

$$Vo = 10\,mV/^{\circ}C$$

EEPROM Memory

Temperature is stored every minute in an ST24C04-type serial I^2C bus compatible EEPROM memory, having a capacity of 512×8 bits, organised as 2 blocks of 256 bytes each. The memory is connected to the microcontroller as an I^2C device where the clock input (SCL) is connected to port RB0 of the microcontroller and the data pin (SDA) is connected to port RB1 of the microcontroller. Although any value pull-up resistors from 1.8 to 47K can be used, in this project 4.7K resistors are used for the I^2C bus. ST24C04 is an 8-pin device with the following pin descriptions:

> Pin 1: No connection
> Pin 2: Device address A1
> Pin 3: Device address A2
> Pin 4: Gnd
> Pin 5: Data line
> Pin 6: Clock line
> Pin 7: Write protect pin
> Pin 8: Vcc.

The device address on the I^2C bus consists of 7 bits

> 4-bit control code
> 2-bit device address (A1 and A2)
> 1-bit block select (if more than one block is used).

Address is sent on the bus as an 8-bit byte where the eighth bit is the R/W control bit. $R/W = 0$ to write to a device, and $R/W = 1$ to read from a device. The 8-bit address format for the ST24C04 consists of the following bits (b is the block-select bit sent by the Master):

1	0	1	0	A2	A1	B	R/W

In this project A1 and A2 inputs are connected to ground so that the memory-select address is hexadecimal \$A0 (bit pattern "10100000") for the first block of memory (256 bytes) when $B = 0$, and \$A2 (bit pattern "10100010") for the other block of memory (256 bytes) when $B = 1$. Note that A1 and A2 are not used by this memory chip (i.e. there are no internal connections to these pins). Write protect pin should be connected to ground to enable writing to the device.

After writing a byte to the memory it is recommended by the manufacturers to wait for about 10 ms before another byte is written or read.

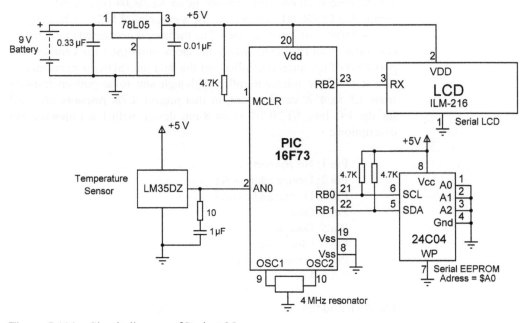

Figure 5.114   Circuit diagram of Project 25

Pin layout of the ST24C04 serial EEPROM memory is shown in Figure 5.115.

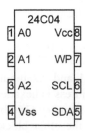

Figure 5.115   Pin layout of ST24C04 memory

Serial LCD

An ILM-216-type serial LCD is used in this project where the serial input of the LCD is connected directly to RB2 port of the microcontroller. The operation of this serial LCD is described in detail in Section 4.3.2. The communication parameters have been selected as: 2400 baud, 8 data bits, and no parity bit.

The operation of the project is very simple: the output of the temperature sensor is converted into digital format every minute and then stored in the EEPROM memory. Data is collected for 1 h (60 samples) and at the end of this time the maximum temperature is found and displayed on the serial LCD.

Flow diagram:

The flow diagram of the project is given in Figure 5.116. At the beginning of the program the A/D converter parameters are defined, and port directions are configured. A loop is used to read the temperature every minute, convert into degrees centigrade and store in the EEPROM memory. The values stored in the EEPROM memory are then read and the maximum value is found and displayed on the LCD.

Software:

PicBasic

I^2C input and output commands by default use the RA0 and RA1 pins for data and clock, respectively. Looking at the A/D configuration of PIC16F73, it is not possible to configure RA0 and RA1 as digital pins and any other pin of PORTA as an analog channel. As a result of this, it is not possible to implement this project using the PicBasic language unless the I^2C routines are modified.

PicBasic Pro

PicBasic Pro program listing of the project is shown in Figure 5.117. At the beginning of the program the A/D converter parameters are defined, and port directions are configured. The A/D converter is then initialised and configured. A *FOR* loop is used where inside this loop the temperature is read from the sensor every minute using the ADCIN statement, it is then converted into degrees centigrade and stored in successive locations of the EEPROM memory using the I2CWRITE statement. The loop is repeated 60 times (i.e. for 1 h) and the loop index (variable *Addr*) is used to address the EEPROM memory. After the data collection another *FOR* loop is used to read the temperature values from the EEPROM (using the I2CREAD statement) memory and then find the largest temperature during the hour. The maximum temperature is stored in variable *Maxone* and is displayed on the LCD in the following format:

$$Max = nnC$$

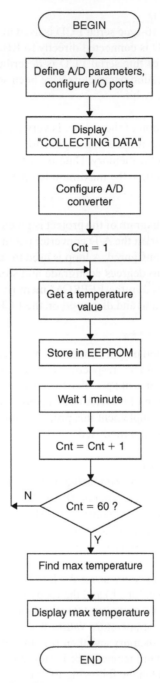

Figure 5.116   Flow diagram of Project 25

Notice that the PicBasic Pro statement SEROUT is used to send serial data to the LCD. RB2 is defined as the serial output port (Sout) and the baud rate is chosen as 2400. The include file "modedefs.bas" contains the definitions for the various PicBasic Pro baud rates. SEROUT command assumes a 4 MHz oscillator when generating its serial bit timing.

Note that the serial data must be inverted before sending to the serial LCD. Mode "N2400" defines the baud rate as 2400 and also inverts the serial output data.

```
'**************************************************************
'
'       SERIAL LCD BASED THERMOMETER WITH SERIAL EEPROM
'       ====================================================
'
' In this project an LCD display is connected to a PIC16F73 microcontroller.
' The microcontroller is configured to operate with a 4 MHz external rezonator.
'
' The project is a thermometer with an external serial EEPROM.
' The temperature is measured every minute and is stored in the EEPROM
' memory. After one hour the measurement stops and the maximum
' temperature during this time is found and displayed on the serial LCD.
'
' A LM35DZ type analog output temperature sensor is used in this project.
' LM35DZ provides an output voltage proportional to the measured temperature.
'
' A 24C04 type serial EEPROm is used in the project.
'
' A serial LCD is used in this project. The Baud rate is selected as 2400.
' The connection between the LCD and the microcontroller is as follows:
'
'       Display         Microcontroller pin
'       ---------       -----------------------
'       RX              RB2
'
' The connection between the microcontroller and the serial EEPROM is as
' follows:
'
'       EEPROM          Microcontroller pin
'       ------------    -----------------------
'       SCL             RB0
'       SDA             RB1
```

Figure 5.117 (Continued)

```
'
' The temperature sensor is connected to the microcontroller as follows:
'
'       Sensor          Microcontroller pin
'       --------         -----------------------
'       LM35DZ          AN0 (RA0)
'
' The PIC16F73 microcontroller has built in 8-bit 5 channel A/D converters.
' The A/D reference voltage is set to +5V. With 8-bit converters, operating with
' a reference voltage of +5V, the bit resolution is 5000/256 = 19.53mV.
'
' The maximum temperature is displayed in degrees C in the following format:
'
'       Max=nnC
'
' In this program PicBasic statement ADCIN is used to read analog data
'
'
' Author:        Dogan Ibrahim
' Date:          November, 2005
' Compiler:      PicBasic Pro
' File:          SERIAL.BAS
'
' Modifications
' ===========
'
'***********************************************************************

        INCLUDE "modedefs.bas"
'
' DEFINITIONS
'
' Define A/D converter parameters
'
DEFINE ADC_BITS          8            ' A/D number of bits
DEFINE ADC_CLOCK         3            ' Use A/D internal RC clock
DEFINE ADC_SAMPLEUS      50           ' Set sampling time in us
'
' Variables used
'
Symbol Sout = 2                       ' RB2 is serial output
Symbol SDA = PORTB.1                  ' EEPROM Data pin
Symbol SCL = PORTB.0                  ' EEPROM clock pin
```

Figure 5.117   (Continued)

```
'
' Variables
'
Res            Var      Byte              ' A/D converter result
Temp1          Var      Byte              ' Temperature in degrees C
Maxone         Var      Byte              ' Maximum temperature
Addr           Var      Byte              ' Address of EEPROM
'
' Start of Program
'
        TRISA = 1                         ' RA0 (AN0) is input
        TRISB = 0                         ' PORTB is output

        PAUSE 500                         ' Wait 0.5sec for LCD to initialize
'
' Clear display and display message "COLLECTING DATA…"
'
        SEROUT Sout, N2400, [12, "COLLECTING DATA…"]
'
' Initialize the A/D converter
'
        ADCON1 = 0                ' Make AN0 to AN4 as analog inputs,
                                  ' make reference voltage = VDD
        ADCON0 = %11000001        ' A/D clock is internal RC, select channel AN0
                                  ' Turn on A/D converter
'
' Start A/D conversion and get 60 samples for an hour
'
        FOR Addr = 0 TO 59
                ADCIN 0, Res                      ' Read Channel 0 data
                Temp1 = 2 * Res                   ' Convert to degrees C
                I2CWRITE SDA, SCL, %10100000,Addr, [Temp1]
                PAUSE 60000                       ' Wait 1 minute
        NEXT Addr                                 ' Repeat
'
' Read all collected data from EEPROM and find and display the largest one.
'
        TRISB = 2                         ' RB1 is input now
        Maxone = 0                        ' Initially maximum = 0
        FOR Addr = 0 TO 59
                I2CRead SDA, SCL, %1010000,Addr, [Temp1]
                IF Temp1 > Maxone THEN Maxone = Temp1
        NEXT Addr
```

Figure 5.117   (Continued)

'

' Max temperature is in variable Maxone.
' Clear display and display the value of Maxone
'

      SEROUT Sout, N2400, [12, "Max = ",#Maxone,"C"]

      END   ' End of program

Figure 5.117   PicBasic Pro listing of Project 25

Project 26

Project title: Programmable thermometer with RS232 serial output

Project description: In this project a programmable digital thermometer is designed and the temperature readings are sent out at required intervals through an RS232 serial line. The project consists of a temperature sensor, a PIC microcontroller and an RS232 line.

The temperature is sent out either in degrees centigrade or in degrees Fahrenheit in the following format:

> nnC
> nnC
> ....
> ....

or,

> nnF
> nnF
> ....
> ....

The thermometer can be connected to a serial line such as the COM1 or COM2 port on a PC. A terminal emulator program such as *Hyperlink*, *SmarTerm*, etc. can be activated on the PC to communicate with the thermometer. The communication parameters should be set to 2400 Baud, 8 data bit, 1 stop bit, and no parity bit. When the thermometer is connected to the PC and the terminal emulation program is activated the following messages will be displayed on the screen. The texts entered by the user are in bold for clarity:

> Digital Thermometer With RS232 Output
> ================================
>
> Enter sampling interval in seconds : **1**
> Output in degrees C (C) or degrees F (F) : **C**
> Press ENTER to start data collection...
>
> Data collection started:
>
> nnC
> nnC
> nnC
> .....
> .....
> .....

Figure 5.118 shows the block diagram of the project where the temperature sensor is connected to one of the analog-to-digital converter (A/D) channels of a PIC microcontroller.

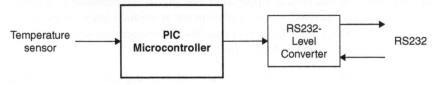

Figure 5.118   Block diagram of Project 26

Hardware: The circuit diagram of the project is shown in Figure 5.119. Any type of PIC microcontroller with a built-in A/D converter can be used. In this project a PIC16F877-type microcontroller is used. This is a popular microcontroller having 40-pins and 8 channel 10-bit multiplexed built-in A/D converter. The reason for choosing this microcontroller is to make your-self familiar with this popular microcontroller.

LM35DZ analog temperature sensor is connected to bit 0 of PORTA (AN0). RB0 and RB1 are configured as RS232 serial output and input, respectively. RS232 voltage levels are $\pm 12V$ where $-12V$ is called Mark (logic 1) and $+ 12$ V is called Space (or logic 0). Normally RS232 voltage

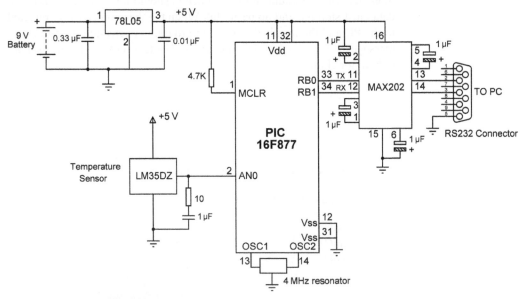

Figure 5.119   Circuit diagram of Project 26

levels are converted to CMOS levels using RS232-level converter chips, such as the MAX202, MAX232, DS275, etc. An RS232-level converter chip converts the 0 to +5 V output from the microcontroller into ±12 V RS232 levels. Similarly, the RS232-level output from a device is converted into 0 to +5 V suitable for the microcontroller inputs.

MAX202 is a 16-pin IC having dual RS232 transmitters and receivers. This IC requires external capacitors for its operation. Figure 5.120 shows the connection diagram when one of the channels of MAX232 is used.

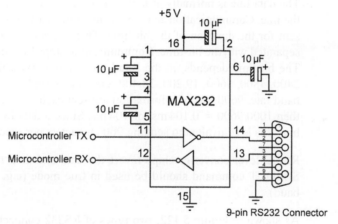

Figure 5.120   MAX232 RS232-level converter

DS275 is a smaller chip with only 8-pins. This IC also includes a transmitter and a receiver. The advantage of DS275 is there is no need to use external capacitors. Figure 5.121 shows the connection diagram when the DS275 is used.

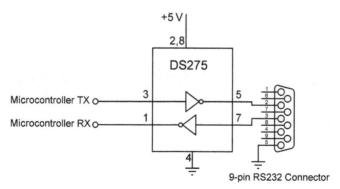

Figure 5.121   DS275 RS232 level converter

In an asynchronous RS232 communication, data is sent and received as frames. A frame consists of a start bit, 7 or 8 data bits, an even or odd parity bit, and a stop bit. In many applications a 10-bit frame is used to send a data byte with the following characteristics:

- 1 start bit
- 8 data bits
- no parity bit
- 1 stop bit.

The data line is normally at logic 1 (MARK) and this is the idle state of the line. Communication starts by sending the start bit which is a logic 0, sent for the duration of the bit-time. Then the 8 data bits are sent, each separated with the bit-time. Communication stops by sending the stop bit. The bit-time depends on the Baud rate chosen. Typical baud rates are: 2400, 4800, 9600, 19,200, 38,400, etc. For example, when using a 9600 baud rate, 9600 bits of information are sent each second. The bit-time is then 1000/9600 = 0.104 ms, or 104 ms. Since a data byte consists of 10 bits, this is equivalent to sending 960 characters per second.

RS232-level converter chips invert the data and as a result of this the SEROUT command should be used in true mode (e.g. T2400 for 2400 baud).

As shown in Figure 5.122, two types of RS232 connector are available: 9-pin D-type, and 25-pin D-type connector. Minimum signals required for RS232 communication are: transmit (TX), receive (RX), and ground. The pin numbers for both types of connectors are

Function	9-way	25-way
TX	2	2
RX	3	3
GND	5	7

 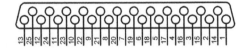

Figure 5.122   RS232 connectors

Flow diagram: The flow diagram of the project is shown in Figure 5.123. At the beginning of the program I/O ports and the A/D are configured. Then the heading is displayed and the user is prompted to enter the sampling interval and the mode as either C (degrees C) or F (degrees F). The program then enters a loop where the temperature is read from the sensor, converted into digital, scaled and then sent to the RS232 port of the microcontroller. The program then waits for the amount of sampling interval and the above process is repeated.

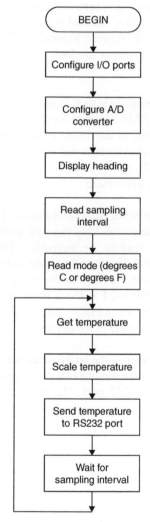

Figure 5.123   Flow diagram of Project 26

Software: **PicBasic**
                       The PicBasic program listing of the project is shown in Figure 5.124. At the beginning of the program addresses of SFR registers used in the

program and the variables are defined. Symbols RS232_out and RS232_in are assigned to 0 and 1, respectively which denote RB0 and RB1. PORTA and PORTB directions are then configured. Notice that when the microcontroller is powered up the RS232 port output may be at logic 0 and this may cause some unwanted data to be sent to the receiving device. In order to avoid this, RS232 port output (RB0) is set to logic 1 for about 100 ms. Then the heading is sent to the RS232 port and the user is prompted to enter the sampling interval and the type of output requested, i.e. degrees Centigrade or degrees Fahrenheit. Serial outputs are sent using the SEROUT statements. Similarly, serial inputs are received using the SERIN statements.

The A/D converter is then initialised and the conversion is started by setting bit 2 of ADCON 0 to logic 1. When the conversion is complete the upper two bits of the 10-bit result is available in register ADRESH and this is copied to variable *Resh*. Similarly, low 8-bits are available in variable ADRESL and is copied to variable *Resl*. Variable *Res* stores the 10-bit result of the conversion.

The A/D converter has a resolution of 10-bits. Thus, it is required to multiply the value read from the A/D converter with 5000/1024 so that we obtain the reading in millivolts. The temperature sensor output is 10 mV/°C and thus, it will be necessary to divide the result by 10 in order to obtain the result in degrees Centigrade. Thus, the required operation is $5000/(1024 \times 10) = 0.48$. In the program, the A/D reading is multiplied by 48 and then divided by 100 to have the final result as true degrees centigrade of temperature.

In the final part of the program the temperature is converted into degrees Fahrenheit if the mode has been selected as "F". The temperature is then sent to the RS232 port. The process repeats after a delay of *TSample* milliseconds.

```
'*********************************************************************
'
'          PROGRAMMABLE THERMOMETER WITH RS232 OUTPUT
'          ==================================================
'
'
' In this project an analog temperature sensor (LM35DZ) is connected to one of
' the A/D channels of a PIC16F877 microcontroller. The microcontroller is
' operated from a 4 MHz external rezonator.
```

Figure 5.124   (Continued)

'
' The thermometer is connected to either COM1 or the COM2 serial port
' of a PC. A terminal emulation program, such as the Hyperterminal is
' activated on the PC to communicate with the thermometer. During this
' communication the user is prompted to enter the sampling interval and the
' mode of the output required (degrees C or degrees F).
'
' A typical communication between the thermometer and the PC is as
' follows (in this example the sampling interval is selected as 2 seconds, and
' the output is requested in degrees C):
'
'   Digital Thermometer With RS232 Output
'   =====================================
'
'   Enter sampling interval in seconds: 2
'   Output is degrees C (C) or degrees F (F) : C
'   Press ENTER to start data collection...
'
'   Data collection started:
'
'   nnC
'   nnC
'   .....
'   .....
'
' PORTB pins RB0 and RB1 are configured as RS232 TX and RX lines respectively.
' RB0 is connected to pin 2 of the RS232 connector. Similarly, RB1 is connected
' to pin 3 of the RS232 connector. The communication parameters are selected as
' follows:
'
'   2400 baud
'   1 start bit
'   8 data bits
'   No parity
'   1 stop bit
'
' The temperature sensor is connected to the microcontroller as follows:
'
'   Sensor   Microcontroller pin
'   -----------   ------------------------
'   LM35DZ   AN0 (RA0)
'
' The PIC16F877 microcontroller has built in 10-bit 8 channel A/D converters.
' The A/D reference voltage is set to +5V.

Figure 5.124   (Continued)

```
'
' In this program PicBasic statement ADCIN is used to read analog data
'
'
' Author:        Dogan Ibrahim
' Date:          November, 2005
' Compiler:      PicBasic
' File:          RS232-1.BAS
'
' Modifications
' ============
'
'*********************************************************************
'
' DEFINITIONS
'
'

Symbol ADCON0 = $1F        ' Address of ADCON0
Symbol ADCON1 = $9F        ' Address of ADCOn1
Symbol ADRESH = $1E        ' Address of ADRESH
Symbol ADRESL = $9E        ' Address of ADRESL
Symbol TRISA = $85         ' Address of TRISA
Symbol TRISB = $86         ' Address of TRISB
Symbol PORTA = $05         ' Address of PORTA
Symbol PORTB = $06         ' Address of PORTB

' VARIABLES
'

Symbol Mode = B1           ' Mode (C or F)
Symbol D = B2
Symbol Dummy = B3
Symbol TSample = W2        ' Sampling time (seconds)
Symbol Resl = W3
Symbol Resh = W4
Symbol Res = W5
Symbol Temp1 = W6          ' Temperature

'
' SYMBOLS
'

Symbol RS232_out = 0       ' RB0 is RS232 output
Symbol RS232_in = 1        ' RB1 is RS232 input
```

Figure 5.124   (Continued)

```
'
' CONSTANTS
'
Symbol CR = 13          ' Carriage-return character
Symbol LF = 10          ' Line-feed character

        POKE TRISA, 1   ' RA0 (AN0) is input
        POKE TRISB, 2   ' RB0=output, RB1=input

        POKE PORTB, 1
        PAUSE 100


'
' Send Heading
'
Again:
        SEROUT RS232_out, T2400, (LF,CR, "Digital Thermometer With RS232 Output")
        SEROUT RS232_out, T2400, (LF,CR, " ================================")
        SEROUT RS232_out, T2400, (LF,LF,CR, "Enter sampling interval in seconds : ")
        SERIN RS232_in, T2400, #TSample
        SEROUT RS232_out, T2400, (#Tsample)
        SEROUT RS232_out, T2400, (LF,CR, "Degrees C (C) or degrees F (F) : ")
        SERIN RS232_in, T2400, Mode
        SEROUT RS232_out, T2400, (Mode)
        SEROUT RS232_out, T2400, (LF,CR, "Press ENTER to start...")
        SERIN RS232_in, T2400, Dummy
        SEROUT RS232_out, T2400, (LF,CR)

        TSample = TSample*1000
'
' Initialize the A/D converter
'
        POKE ADCON1, %10001110    ' Make AN0 analog input,
                                  ' make reference voltage = VDD
        POKE ADCON0, %01000001    ' A/D clock is internal, select channel AN0
                                  ' Turn on A/D converter
More:
' Start A/D conversion and get 60 samples for an hour
```

Figure 5.124 (Continued)

'
'
```
           D = "C"

           PEEK ADCON0, B0
           Bit2 = 1
           POKE ADCON0, B0                  ' Start A/D conversion

WT:        Pause 1
           PEEK ADCON0, B0
           IF Bit2 = 1 THEN WT

           PEEK ADRESH, Resh                ' Get high byte
           PEEK ADRESL, Resl                ' Get low byte
           Res = Resh*256 + Resl

           Temp1 = 48 * Res                 ' Convert to degrees C
           Temp1 = Temp1/100
           IF Mode = "C" THEN Cent          ' If Fahrenheit
           Temp1 = Temp1 * 18
           Temp1 = Temp1 + 320
           Temp1 = Temp1 / 10
           D = "F"

Cent:
           SEROUT RS232_out, T2400, (LF,CR, #Temp1, D)
           PAUSE TSample
           GOTO More

           END                              ' End of program
```

Figure 5.124   PicBasic listing of Project 26

PicBasic Pro
The PicBasic Pro program listing of the project is shown in Figure 5.125. At the beginning of the program the A/D parameters are defined. Symbol *RS232_out* and *RS232_in* are defined as the RS232 output and input ports, respectively.

The main program starts with label *Again* where the heading text is sent to the RS232 port. Then the user is requested to enter the sampling interval in seconds. The received value is stored in variable *TSample*. If the user does not enter any characters in 5 s (5000 ms), the SERIN input routine times out and program jumps to label *ESample*, where the input is requested again. Similarly, the user is requested to enter the output mode as either degrees C or as degrees F. The required mode of temperature is stored in variable *Mode*. If the user does not enter any characters in 5 s, the SERIN input routine times out and jumps to label *EMode*.

The AD converter is then initialised by configuring registers ADCON1 and ADCON0. A/D conversion is started by the ADCIN instruction. The A/D converter has a resolution of 10-bits. Thus, it is required to multiply the value read from the A/D converter with 5000/1024 so that we obtain the reading in millivolts. The sensor output is 10 mV/°C and thus, it will be necessary to divide the result by 10 in order to obtain the result in degrees Centigrade. Thus, the required operation is 5000/ (1024 × 10) = 0.48. In the program the A/D reading is multiplied by 48 and then divided by 100 to have the final result as true degrees centigrade of temperature.

In the final part of the program the temperature is converted into degrees Fahrenheit if the mode has been selected as "F". The temperature is then sent to the RS232 port. The process repeats after a delay of *TSample* milliseconds.

```
'******************************************************************
'
'        PROGRAMMABLE THERMOMETER WITH RS232 OUTPUT
'        ====================================================
'
' In this project an analog temperature sensor (LM35DZ) is connected to one of
' the A/D channels of a PIC16F877 microcontroller. The microcontroller is
' operated from a 4 MHz external rezonator.
'
' The thermometer is connected to either COM1 or the COM2 serial port
' of a PC. A terminal emulation program, such as the Hyperterminal is
' activated on the PC to communicate with the thermometer. During this
' communication the user is prompted to enter the sampling interval and the
' mode of the output required (degrees C or degrees F).
'
' A typical communication between the thermometer and the PC is as
' follows (in this example the sampling interval is selected as 2 seconds, and
' the output is requested in degrees C):
```

Figure 5.125   (Continued)

```
'         Digital Thermometer With RS232 Output
'         ==================================
'
'         Enter sampling interval in seconds: 2
'         Output is degrees C (C) or degrees F (F) : C
'         Press ENTER to start data collection…
'
'         Data collection started:
'
'         nnC
'         nnC
'         .....
'         .....

' PORTB pins RB0 and RB1 are configured as RS232 TX and RX lines respectively.
' RB0 is connected to pin 2 of the RS232 connector. Similarly, RB1 is connected
' to pin 3 of the RS232 connector. The communication parameters are selected as
' follows:
'
'         2400 baud
'         1 start bit
'         8 data bits
'         No parity
'         1 stop bit
'
' The temperature sensor is connected to the microcontroller as follows:
'
'         Sensor          Microcontroller pin
'         -----------     ------------------------
'         LM35DZ          AN0 (RA0)
'
' The PIC16F877 microcontroller has built in 10-bit 8 channel A/D converters.
' The A/D reference voltage is set to +5V.
'
' In this program PicBasic statement ADCIN is used to read analog data
'
'
' Author:          Dogan Ibrahim
' Date:            November, 2005
' Compiler:        PicBasic Pro
' File:            RS232-2.BAS
```

Figure 5.125   (Continued)

```
'
' Modifications
' ==========
'

'*********************************************************************
          INCLUDE "modedefs.bas"

'

' DEFINITIONS
'

' Define A/D converter parameters
'

DEFINE ADC_BITS          10          ' A/D number of bits
DEFINE ADC_CLOCK         3           ' Use A/D internal RC clock
DEFINE ADC_SAMPLEUS      50          ' Set sampling time in us
'

' VARIABLES
'

Tsample   VAR Word               ' Sampling time (seconds)
Mode      VAR Byte               ' Mode (C or F)
Dummy     VAR Byte
D         VAR Byte               ' Temperature mode display


'

' SYMBOLS
'

Symbol RS232_out = 0             ' RB0 is RS232 output
Symbol RS232_in = 1             ' RB1 is RS232 input
'

' CONSTANTS
'

CR        CON   13               ' Carriage-return character
LF        CON   10               ' Line-feed character
'

' Variables
'

Res      Var    Word            ' A/D converter result
Temp1    Var    Word            ' Temperature in degrees C

         TRISA = 1              ' RA0 (AN0) is input
         TRISB = 2              ' RB0 = output, RB1 = input
         PAUSE 1000
```

Figure 5.125   (Continued)

```
'
' Send Heading to RS232 port
'


Again:
        SEROUT RS232_out, T2400, [LF,CR, "Digital Thermometer With RS232 Output"]
        SEROUT RS232_out, T2400, [LF,CR, " ================================="]
Esample:
        SEROUT RS232_out, T2400, [LF,LF,CR, "Enter sampling interval in seconds : "]
        SERIN RS232_in, T2400, 5000, ESample, #TSample
        SEROUT RS232_out, T2400, [#Tsample]
EMode:
        SEROUT RS232_out, T2400, [LF,CR, "Degrees C (C) or degrees F (F) : "]
        SERIN RS232_in, T2400, 5000, EMode, Mode
        SEROUT RS232_out, T2400, [Mode]
Estart:
        SEROUT RS232_out, T2400, [LF,CR, "Press ENTER to start..."]
        SERIN RS232_in, T2400, 5000, Estart, Dummy
        SEROUT RS232_out, T2400, [LF,CR]


        TSample = TSample*1000                  ' Convert to ms
'
' Initialize the A/D converter
'

        ADCON1 = %10001110                      ' Make AN0 analog inputs,
                                                ' Reference voltage = VDD
        ADCON0 = %01000001                      ' A/D clock is internal, Select channel AN0
                                                ' Turn on A/D converter


More:
' Start A/D conversion and get 60 samples for an hour
'
'

        D = "C"
        ADCIN 0, Res                            ' Read Channel 0 data
'
' Scale the reading to obtain degrees C. This involves multiplying by
' 5000/1024 and then diviing to 10 since the sensor output is 10 mV/C. i.e.
' We have to multiply the A/D readings with 5000/(1024 × 10) which
' is equal to 0.48. We thus multiply by 48 and then divide by 100
'

        Temp1 = 48 * Res                        ' Convert to degrees C
        Temp1 = Temp1/100
```

Figure 5.125   (Continued)

'

' If the required output is degrees Fahrenheit, we have to perform the
' operation: 1.8C + 32. Here, we are multiplying by 10. ie. Multiply by 18 and
' add 320. The final result is then divided by 10.

'

```
        IF Mode = "F" THEN                    ' If Fahrenheit selected
                Temp1 = Temp1 * 18
                Temp1 = Temp1 + 320
                Temp1 = Temp1 / 10
                D = "F"
        ENDIF
```

'

' Send temperature to RS232 port, wait for sampling time and repeat

'

```
        SEROUT RS232_out, T2400, [LF,CR, #Temp1, D]
        PAUSE Tsample
        GOTO More

        END     ' End of program
```

Figure 5.125 PicBasic Pro listing of Project 26

Figure 5.126 shows a sample output obtained when the *SmartTerm* terminal emulation program is used (we can use any type of terminal emulation software) to communicate with the thermometer. In this example, the sampling interval is selected as 4 s and the output is requested as degrees C.

The project built on a breadboard is shown in Figure 5.127.

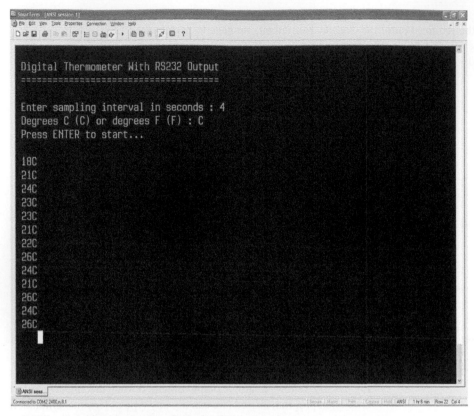

Figure 5.126   Sample output taken from the PC screen

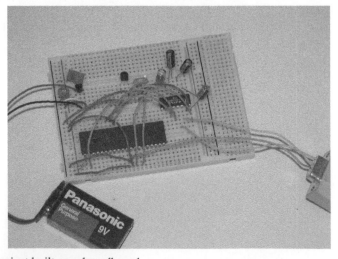

Figure 5.127   Project built on a breadboard

Project 27

Project title: Electronic organ

Project description: This is a simple electronic organ project. A small speaker is connected to PORTA of a PIC microcontroller. Eight push-button switches are connected to PORTB to act as the keyboard for the electronic organ. Only one octave (eight notes) is provided in this project.

Figure 5.128 shows the block diagram of the project.

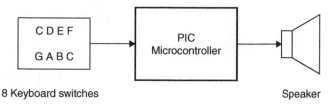

Figure 5.128 Block diagram of Project 27

Hardware: The circuit diagram of the project is shown in Figure 5.129. Although any model of PIC microcontroller with at least 9 I/O pins can be used, a PIC16F627 microcontroller is used in this project. The microcontroller is operated from an external 4 MHz resonator. A small speaker is connected to bit 0 of PORTA (RA0) using a 10 μF electrolytic capacitor.

Keyboard switches are connected to PORTB. Bit 0 is assigned to musical note C, bit 1 is assigned to note D, bit 2 is assigned to note E, and so on. The switches are normally held at logic HIGH using the internal PORTB pull-up resistors. Pressing a switch sends a logic LOW to the corresponding microcontroller input port pin.

In this project, the following octave of notes is used:

Switch	1	2	3	4	5	6	7	8
Note	C	D	E	F	G	A	B	C
Frequency	262	294	330	349	392	440	494	524

The program continuously checks the switches and if any switch is pressed then the musical note corresponding to that switch position is sent to the speaker.

Flow diagram: The flow diagram of the project is shown in Figure 5.130. At the beginning of the project PORTA and PORTB directions are configured and

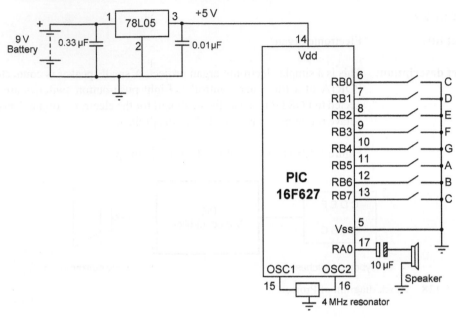

Figure 5.129   Circuit diagram of the project

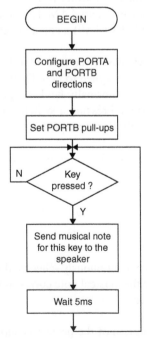

Figure 5.130   Flow diagram of Project 27

PORTB internal pull-up resistors are enabled. The program then enters an endless loop where the switches are checked. If a switch is pressed, then the musical note corresponding to that switch position is sent to the speaker. The program waits for 5 ms and then the above process is repeated.

Software:

PicBasic

PicBasic language does not have an instruction to generate a signal with the required frequency. A signal with a required frequency can be generated using the timer interrupt. But unfortunately, PicBasic language does not support the use of interrupts from a high-level language. As a result of this, it is not very easy to generate musical notes from the PicBasic language. Only the PicBasic Pro program of this project is given here.

PicBasic Pro

The PicBasic Pro program listing of the project is shown in Figure 5.131. At the beginning of the program the frequencies of musical notes are stored in an array called Notes. Then, PORTB is configured as input and PORTA is configured as output. PORTB internal pull-up resistors are then enabled so that the switches are normally held at logic HIGH. The statement IF PORTB <> 255 is true if any switch is pressed. The status of PORTB is then inverted and the bit which is 0 is the bit position pressed by the user. For example, if the user pressed switch 5, number 16 will be obtained.

Normal state of PORTB	1 1 1 1 1 1 1 1
State when key 4 is pressed	1 1 1 0 1 1 1 1
State when PORTB inverted	0 0 0 1 0 0 0 0

PicBasic Pro statement NCD is used to obtain the bit position of the switch pressed. In the above example, if p is the state of PORTB when inverted, then,

$$y = \text{NCD}\, p$$

will return 5 in variable y, i.e. bit position 5 is set in variable p. Thus, the statement

$$\text{Key\_pressed} = \text{NCD Key}$$

returns the switch number (1 to 8) pressed. This number is then used as an index in array Notes and the PicBasic Pro statement FREQOUT is used to send the frequency of the required note to the speaker. The note is sounded for a duration of 5 ms.

```
'***********************************************************************
'
'                          SIMPLE ELECTRONIC ORGAN
'                          ================================
'
' In this project a small speaker is connected to bit 0 of PORTA of a PIC16F627
' microcontroller. Also, 8 push-button switches are connected to PORTB
' of the microcontroller. The switches are used to represent the musical notes
' C to C (i.e. one octave). The switch assignments are as follows:
'
'          Switch      Musical note
'          --------     ----------------
'
'          RB0           C
'          RB1           D
'          RB2           E
'          RB3           F
'          RB4           G
'          RB5           A
'          RB6           B
'          RB7           C
'
' The frequencies of the notes used are as follows:
'
'          Note        Frequency (Hz)
'          ------       --------------------
'
'          C           262
'          D           294
'          E           330
'          F           349
'          G           392
'          A           440
'          B           494
'          C           524
'
'
' When a switch is pressed, the frequency of the musical note corresponding
' to that switch is sent to the speaker.
'
' The project can be used to play simple tunes.
```

Figure 5.131   (Continued)

```
'
'
' Author:          Dogan Ibrahim
' Date:            December, 2005
' Compiler:        PicBasic Pro
' File:            SOUND1.BAS
'
' Modifications
' ============
'
'*********************************************************************

' DEFINITIONS
'

Speaker        VAR    PORTA.0            ' Speaker is connected to RA0
Notes          VAR    Word[9]            ' Frequencies of musical notes
Key            VAR    Byte
Key_pressed    VAR    Byte               ' Key pressed


' Define the frequencies of musical notes
'

Notes[1] = 262 : Notes[2] = 294        : Notes[3] = 330        : Notes[4] = 349
Notes[5] = 392 : Notes[6] = 440        : Notes[7] = 494        : Notes[8] = 524
'

' Configure PORT directions
'
       TRISB = %11111111                       ' PORTB is input (keys)
       TRISA = 0                               ' PORTA (RA0) is output
       OPTION_REG.7 = 0                         ' Enable internal PORTB pull-ups
       CMCON = 7                                ' Make RA0 digital I/0
'

' Check if any key is pressed, and if so, find the musical note corresponding
' to the pressed key and send the frequency of this note to the speaker.
'

Loop:
           IF PORTB <> 255       THEN           ' Check if any key pressed
           Key = ~PORTB                         ' Invert key pattern
           Key_pressed = NCD Key                ' Get key pressed
           FREQOUT Speaker,5,Notes[Key_pressed] ' Send note to speaker
       ENDIF

       GOTO Loop                                ' Repeat

       END                                      ' End of program
```

Figure 5.131 PicBasic Pro listing of the project

Improving the musical tones

The tones generated by the statement FREQOUT are square wave and they are very noisy. One way to improve the quality of these tones is by filtering the output of the microcontroller signal. Figure 5.132 shows a simple filter that can be used to obtain a cleaner waveform when the FREQOUT statement is used.

In many applications, the amplitude of the output signal may not be adequate and it may be necessary to amplify this signal. Figure 5.133 shows an amplifier circuit which can be used to increase the output signal level of our electronic organ.

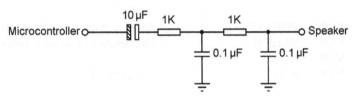

Figure 5.132   A simple filter

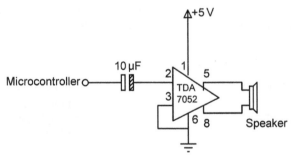

Figure 5.133   Amplifying the output signal

It is also recommended to use a higher oscillator frequency, e.g. 20 MHz for an improved output response. This will require the use of a 20 MHz crystal, and a PIC chip which can operate at 20 MHz. The following line of code should also be added to the program to show that we are using a 20 MHz crystal, and not the default 4 MHz.

DEFINE OSC 20

Project 28

Project title: Unipolar stepping motor control

Project description: This project is about the control of an unipolar stepping motor using a PIC microcontroller. The project shows how a stepping motor can be controlled to rotate clockwise for the required number of revolutions.

In this project the stepping motor is controlled as follows:

> Rotate 100 revolutions clockwise
> Stop

Figure 5.134 shows the block diagram of the project. Four output ports of the microcontroller are connected to MOSFET transistors which drive the stepping motor.

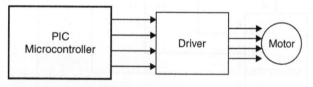

Figure 5.134   Block diagram of Project 28

Hardware: The circuit diagram of the project is shown in Figure 5.135. In this project a PIC16F627-type microcontroller, operated with its internal 4 MHz clock is used. The master clear circuit is enabled during programming of the chip. The stepping motor used in the project is the model UAG2 (see Figure 5.136), manufactured by *SAIA Schrittmotoren*. This stepping motor operates with 12 V, has 6 leads, and a stepping angle of 18°. Thus, 20 steps are required for a complete revolution. The motor consists of two windings and the pin connections are as follows:

Pin	Function
1	Start of first winding
2	Start of second winding
3	Common point of first winding
4	Common point of second winding
5	End of first winding
6	End of second winding

PORTB pins RB0-RB3 are connected to Gate inputs of four IRL1520N type MOSFET power transistors which are used as switches. The Drain outputs of these transistors are connected to motor windings as shown in Figure 5.135. Common points of both windings are connected to +12V supply using 68 Ω current-limiting resistors.

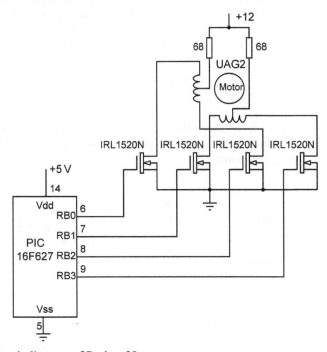

Figure 5.135   Circuit diagram of Project 28

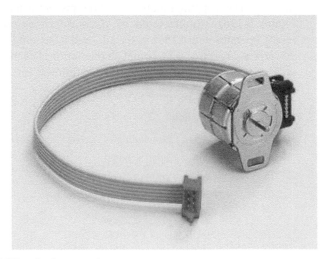

Figure 5.136   UAG2 unipolar stepping motor

Flow diagram: The flow diagram of the project is shown in Figure 5.137. At the beginning of the project PORTB pins are configures as output. Pulses are then sent to PORTB to rotate the motor 100 steps clockwise. The motor is then stopped.

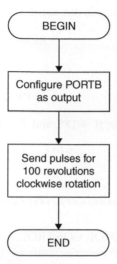

Figure 5.137   Flow diagram of Project 28

Software: **PicBasic**

PicBasic program listing of the project is given in Figure 5.138. At the beginning of the program PORTB pins are configured as output. Variable *Revolutions* stores the required number of revolutions which is 100 in this example. Variable *Pulses* stores the number of pulses to be sent to the motor. This variable is divided by 4 so that it stores the number of times the patterns of 1,2,4,8 are to be sent to the motor. A FOR loop is used to send the pulses to the motor. Pulses are sent as in the following order:

```
......
......
0001
0010
0100
1000
0001
......
......
```

Notice that a 3 ms delay is used between each step output to the motor. The RPM (number of revolutions per minute) of the motor can be calculated as follows:

If T is the time between the steps, and β is the step angle of the motor, then the motor rotates β/T steps in 1 s. Since one revolution is 360°, the number of revolutions in one second is $\beta/360T$. The RPM is then given by

$$RPM = 60\beta/360T$$

or,

$$RPM = \beta/6T$$

In this example, $\beta = 18°$, and $T = 3$ ms (0.003 s). Thus,

$$RPM = 18/6(0.003) = 1000$$

```
'****************************************************************
'
'            UNIPOLAR STEPPING MOTOR CONTROL
'`           =====================================
'
' In this project an UAG2 type unipolar stepping motor is connected to pins
' RB0-RB3 of PORTB of a PIC16F627 microcontroller. The microcontroller is
' operated from its internal 4MHz clock.
'
' The motor is operated as follows:
'
'            Turn motor 100 revolutions clockwise
'            Stop
'
' Four IRL1520N type MOSFET power transistors are used as switches to
' provide current to the motor.
'
' Author:      Dogan Ibrahim
' Date:        December, 2005
' Compiler:    PicBasic
' File:        MOTOR1.BAS
'
' Modifications
' ==========
'
'****************************************************************
```

Figure 5.138 (Continued)

```
'
' Symbols
'
Symbol PORTB = $06                    ' PORTB address
Symbol TRISB = $86                    ' TRISB address

Symbol Revolutions = W0               ' Required number of revolutions
Symbol Pulses = W1                    ' Number of pulses to be sent
Symbol J = B4                         ' Used in FOR loop

        POKE TRISB, 0                 ' PORTB is output

        Revolutions = 100            ' Required number of revolutions
        Pulses = 20*Revolutions      ' Required number of pulses
        Pulses = Pulses / 4          ' Required number of steps
'
' Send Pulses to the motor for clockwise rotation. The number of revolutions is equal
' to Revolutions (100 in this example)
'
        FOR J = 1 TO Pulses
                POKE PORTB, 1
                Pause 3
                POKE PORTB, 2
                Pause 3
                POKE PORTB, 4
                Pause 3
                POKE PORTB, 8
                Pause 3
        NEXT J

        END                           ' End of program
```

Figure 5.138   PicBasic listing of Project 28

PicBasic Pro

PicBasic Pro program listing of the project is given in Figure 5.139. At the beginning of the program TRISB is cleared to zero so that all PORTB pins are configured as outputs. Variable *Steps* is defined as a byte array and this array stores the bit patterns to be sent to the motor for clockwise rotation. For example, sending the bit pattern …,1,2,4,8, … rotates the motor clockwise by 4 steps. Variable *Revolutions* stores the required number of revolutions which is 100 in this example. Variable *Pulses* stores the number of pulses to be sent to the motor. This variable is divided by 4 so that it stores the number of times the patterns of 1,2,4,8 are to be sent to the motor so that the motor rotates clockwise required number of revolutions. Two FOR loops are used in the program. The outer loop controls the number of steps to be sent, and the inner loop sends the bit patterns of 1,2,4,8 to the motor, as in the PicBasic program, the motor rotates with a speed of RPM = 1000.

```
'*********************************************************************
'
'             UNIPOLAR STEPPING MOTOR CONTROL
'             ==================================
'
' In this project an UAG2 type unipolar stepping motor is connected to pins
' RB0-RB3 of PORTB of a PIC16F627 microcontroller. The microcontroller is
' operated from its internal 4MHz clock.
'
' The motor is operated as follows:
'
'         Turn motor 100 revolutions clockwise
'         Stop
'
' Four IRL1520N type MOSFET power transistors are used as switches to
' provide current to the motor.
'
' Author:        Dogan Ibrahim
' Date:          December, 2005
' Compiler:      PicBasic Pro
' File:          MOTOR2.BAS
'
' Modifications
' =============
'
'*********************************************************************
'
' Variables
'
Steps         Var      Byte[4]        ' Step bit patterns
Revolutions   Var      Word           ' Required number of revolutions
Pulses        Var      Word           ' Number of pulses to be sent
I             Var      Byte           ' Used in FOR loop
J             Var      Word           ' Used in FOR loop

        TRISB = 0                     ' PORTB is output
```

Figure 5.139   (Continued)

'

' Define data to be sent to the motor

'

```
            Steps[0] = 1
            Steps[1] = 2
            Steps[2] = 4
            Steps[3] = 8

            Revolutions = 100          ' Required number of revolutions
            Pulses = 20*Revolutions    ' Required number of pulses
            Pulses = Pulses / 4        ' Required number of steps
```

'

' Send Pulses to the motor for clockwise rotation. The number of revolutions is equal
' to Revolutions (100 in this example)

'

```
            FOR J = 1 TO Pulses
                    FOR I = 0 TO 3
                            PORTB = Steps[I]
                            PAUSE 3
                    NEXT I
            NEXT J

            STOP

            END                        ' End of program
```

Figure 5.139   PicBasic Pro listing of Project 28

Project 29

Project title: Unipolar stepping motor control using UCN5804B

Project description: This project is similar to Project 28, but here the stepping motor is controlled using a UCN5804B type motor controller IC. In this project the motor is rotated continuously. Motor direction is controlled using a button. Normally the motor rotates in one direction, and when the button is pressed the direction is reversed.

Figure 5.140 shows the block diagram of the project.

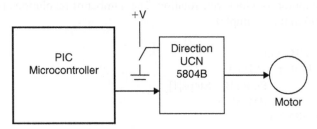

Figure 5.140 Block diagram of Project 29

Hardware: The circuit diagram of the project is shown in Figure 5.141. In this project a PIC16F627-type microcontroller, operated with its internal 4-MHz clock is used. Same stepping motor as in Project 28 is used. RB0 port of the microcontroller is connected to STEP input of the UCN5804B. Direction of the motor is controlled from a button connected to the DIR input. OutA, OutB, OutC, and OutD outputs of the IC are connected to the windings of the motor. KaC and KbD are the common outputs connected to the common points of the motor windings. Motor is rotated by one step each time a pulse is applied to the STEP input of the IC.

Flow diagram: The flow diagram of the project is shown in Figure 5.142. The operation of the project is very simple. After PORTB is configured as output, pulses are sent to UCN5804B continuously with 3 ms delay between each output. As in the previous project, the speed of rotation is 1000 RPM.

Software: **PicBasic**
PicBasic program listing of the project is given in Figure 5.143. At the beginning of the program PORTB is configured as output. Pulses are then sent to RB0 with 3ms delay between each output.

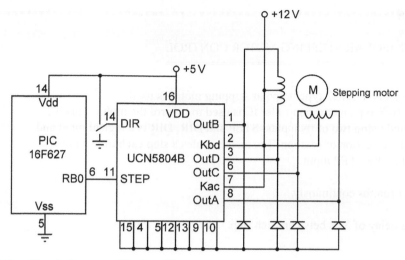

Figure 5.141 Circuit diagram of Project 29

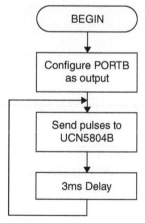

Figure 5.142 Flow diagram of Project 29

PicBasic Pro

PicBasic Pro program listing of the project is given in Figure 5.144. The project is very simple. At the beginning of the Project PORTB is configured as output. Pulses are then sent to port pin RB0 with 3 ms delay between each output.

```
'***************************************************************
'
'            UNIPOLAR STEPPING MOTOR CONTROL
'            =================================
'
' In this project an UAG2 type unipolar stepping motor is used.
' A UCN5804B type motor controller IC is used to control the motor. This IC
' is controlled using two of its inputs: STEP and DIR. DIR is a logical input and
' controls the direction of rotation. The motor rotates a step each time a pulse
' is applied to the STEP input.
'
' The motor rotates continuosly.
'
' There is a delay of 3ms between each step.
'
'
'
' Author:        Dogan Ibrahim
' Date:          January, 2005
' Compiler:      PicBasic
' File:          MOTOR3.BAS
'
' Modifications
' =============
'
'***************************************************************
'
' Symbols
'
Symbol PORTB = $06              ' PORTB address
Symbol TRISB = $86              ' TRISB address

        POKE TRISB, 0           ' PORTB is output

        POKE PORTB, 0           ' Clear STEP to start with

More:
        POKE PORTB, 1           ' Set STEP = 1
        POKE PORTB, 0           ' Set STEP = 0
        PAUSE 3                 ' Wait 3ms
        GOTO More               ' Repeat

        END                     ' End of program
```

Figure 5.143   PicBasic listing of Project 29

```
'*********************************************************************
'
'              UNIPOLAR STEPPING MOTOR CONTROL
'              =======================================
'
' In this project an UAG2 type unipolar stepping motor is used.
' A UCN5804B type motor controller IC is used to control the motor. This IC
' is controlled using two of its inputs: STEP and DIR. DIR is a logical input and
' controls the direction of rotation. The motor rotates a step each time a pulse
' is applied to the STEP input.
'
' The motor rotates continuosly.
'
' There is a delay of 3ms between each step.
'
'
'
' Author:         Dogan Ibrahim
' Date:           January, 2005
' Compiler:       PicBasic Pro
' File:           MOTOR4.BAS
'
' Modifications
' =============
'
'*********************************************************************
'
' Variables
'
Step_input Var PORTB.0              ' Assign Step_input to RB0

        TRISB = 0                   ' PORTB is output

        Step_input = 0              ' Clear STEP to start with
More:
        Step_input = 1              ' Set STEP = 1
        Step_input = 0              ' Set STEP = 0
        Pause 3                     ' Wait 3ms
        GOTO More                   ' Repeat

        END                         ' End of program
```

Figure 5.144 PicBasic Pro listing of Project 29

Project 30

Project title: Servomotor-based mobile robot control

Project description: Mobile robots are used in many industrial, commercial, research, and hobby applications. This project is about the control of a mobile robot using servomotors. The robot used in this project is the base of a popular mobile robot known as *Boe Bot*, developed by *Parallax* (www.parallax.com and www.stampinclass.com). The basic robot is controlled from a *Basic Stamp* controller (Trademark of *Parallax Inc.*). The robot base and electronic circuit have been modified by the author so that the robot can be used with PIC microcontrollers (see Figure 5.145).

The robot consists of two side drive wheels and a caster wheel at the back. The drive wheels are connected to servomotors. A breadboard is placed on the robot base for the electronic control circuit. The robot is driven from a 9V battery, and a 78L05-type voltage regulator is used to obtain +5V to supply power to the microcontroller circuit.

In this project programs are developed to move the robot forward, backward, and to turn left and right.

Figure 5.145   Robot used in the project

Hardware: The circuit diagram of the project is shown in Figure 5.146. In this project a PIC16F84 microcontroller is used and the microcontroller is operated with a 4 MHz crystal.

Servomotors are used to drive the left wheel and the right wheel. A servomotor has three leads: power supply, ground, and the signal pin. Left servomotor is connected to bit 0 of PORTB (RB0), and right servomotor is connected to bit 1 of PORTB (RB1). Although some servomotors can operate with +5V supply, most servomotors require 6–9V to operate.

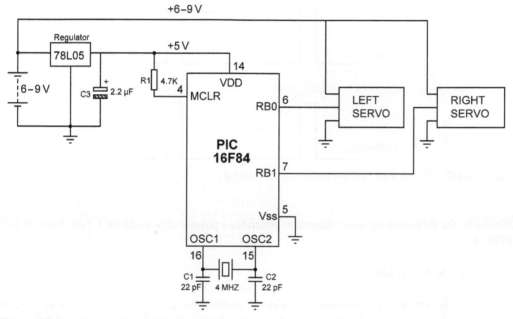

Figure 5.146 Circuit diagram of Project 30

Operating the servomotor

As described in Section 4.7 the servomotors used in robotic applications are modified servos where the motor can rotate in either direction continuously by applying pulses to the servomotor.

In a modified servomotor typically a pulse with a width of 1.3 ms rotates the motor clockwise at full speed. A pulse with a width of 1.7 ms rotates the motor anti-clockwise, and a pulse with a width of 1.5 ms stops the motor. Figure 5.147 shows typical pulses used to drive modified servomotors.

The pulse required to operate a servomotor can very easily be obtained using the PULSOUT statement of the PicBasic and PicBasic Pro compilers. When a 4 MHz crystal is used, the time interval of PULSOUT is in units of 10 μs. For example, the following PicBasic statement generates a pulse with a width of 1.3 ms from bit 0 of PortB (1.3 ms = 1300 μs and 1300/10 = 130):

PULSOUT 0, 130

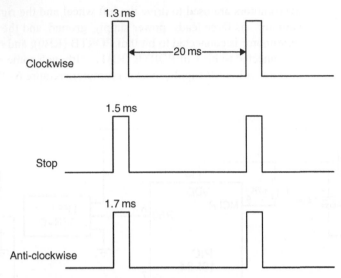

Figure 5.147   Pulses used to drive modified servomotors

Similarly, the following PicBasic statement generates a pulse with a width of 1.7 ms from bit 1 of PORTB:

    PULSOUT 1, 170

A single pulse rotates the servomotor by a small amount. For a continuous rotation we have to apply the pulses continuously. In most applications a loop is formed in software and pulses are sent to the servomotor continuously. A delay is inserted between each pulse. The duration of this delay determines the speed of the motor and about 20 ms is most commonly used value.

The following PicBasic (or PicBasic Pro) code shows how a servomotor connected to port RB0 can be rotated clockwise continuously:

```
Loop:        PULSOUT 0, 130        ' Send a pulse
             PAUSE 20              ' Wait 20 ms
             GOTO Loop             ' Repeat
```

Similarly, the following PicBasic (or PicBasic Pro) code shows how a servomotor connected to port RB1 can be rotated anti-clockwise continuously:

```
Loop:        PULSOUT 1, 170        ' Send a pulse
             PAUSE 20              ' Wait 20 ms
             GOTO Loop             ' Repeat
```

You can experiment by varying the pulse width and the delay to see how the speed of the motor changes.

Forward movement

Assuming that two side wheels are connected to servomotors, the robot moves forward when

> Left wheel rotates anti-clockwise
> Right wheel rotates clockwise

In this project, the left servomotor is connected to port pin RB0 and right servomotor is connected to port pin RB1. The following PicBasic (or PicBasic Pro) code can then be used to move the robot forward:

```
Forward:     PULSOUT 0, 170       ' Left wheel anti-clockwise
             PULSOUT 1, 130       ' Right wheel clockwise
             PAUSE 20             ' Wait 20 ms
             GOTO Forward         ' Repeat
```

Backward movement

Assuming that the two side wheels are connected to servomotors, the robot moves backward when

> Left wheel rotates clockwise
> Right wheel rotates anti-clockwise

In this project, the left servomotor is connected to port pin RB0 and right servomotor is connected to port pin RB1. The following PicBasic (or PicBasic Pro) code can then be used to move the robot backward:

```
Backward:    PULSOUT 0, 130       ' Left wheel clockwise
             PULSOUT 1, 170       ' Right wheel anti-clockwise
             PAUSE 20             ' Wait 20 ms
             GOTO Backward        ' Repeat
```

Moving the robot for required amount of time

The code given above moves the robot forward or backward continuously. There are applications where we may want to mode the robot only required amount of time. For example, we may want to move the robot forward for 5 s, then stop for 3 s, and then move backward for 2 s.

We can adjust the movement time by using a FOR loop. The following code shows how we can move the robot forward using a FOR loop:

```
FOR J = 1 TO M
        PULSOUT 0, 170
        PULSOUT 1, 130
        PULSOUT 20
NEXT J
```

In this code variable M determines the number of times the loop is executed. Ignoring the small time taken by the FOR and the NEXT statements, the time taken to execute only one iteration of the FOR loop can be determined approximately as

```
FOR J = 1 TO M
        PULSOUT 0, 170              1.7 ms
        PULSOUT 1, 130              1.3 ms
        PULSOUT 20                  20.0 ms
NEXT J                             ----------
                                    23.0 ms
```

Thus, if the robot is required to move for T seconds ($1000 \times T$ ms) forward or backward, the value of M to be used in the FOR loop can be calculated as follows:

$$M = 1000 \times T/23$$

An example is given below.

Example 1

A mobile robot is controlled with two servomotors as shown in Figure 5.146. Write a PicBasic program which will perform the following operations:

Move the robot forward for 4 s
Wait for 5 s
Move the robot backward for 3 s
Stop

Solution 1

The first action is to move the robot forward for 4 s. Thus, the value of M is

$$M = 4000/23 = 174$$

Then the robot is required to stop for 5 s and then move backward for 3 s. The value of M for this movement is

$$M = 3000/23 = 130$$

The program is very simple and consists of only a few lines.

PicBasic program for this example is given in Figure 5.148.

```
'****************************************************************
'
'                        ROBOT CONTROL
'                        ================
'
' In this project a mobile robot is controlled. The robot has two side wheels and
' a back caster wheel. Side wheels are connected to servomotors as follows:
'
'        Left wheel        RB0
'        Right wheel       RB1
'
' In this project the robot moves as follows:
'
'        Move the robot forward for 4 seconds
'        Wait for 5 seconds
'        Move the robot backward for 3 seconds
'        Stop
'
' A PIC16F84 type microcontroller is used with a 4 MHz crystal
'
' Author:        Dogan Ibrahim
' Date:          January, 2005
' Compiler:      PicBasic
' File:          SERVO1.BAS
'
' Modifications
' ===========
'
'****************************************************************
```

Figure 5.148   (Continued)

```
'
' Symbols
'

Symbol PORTB = $06              ' PORTB address
Symbol TRISB = $86              ' TRISB address
Symbol J = B0

        POKE TRISB, 0           ' PORTB is output

'
' Move the robot forward for 4 seconds
'
        FOR J = 1 TO 174
                PULSOUT 0, 170
                PULSOUT 1, 130
                PAUSE 20
        NEXT
'
' Wait for 5 seconds
'
        PAUSE 5000
'
' Move the robot backward for 3 seconds
'
        FOR J = 1 TO 130
                PULSOUT 0, 130
                PULSOUT 1, 170
                PAUSE 20
        NEXT J

        END                     ' End of program
```

Figure 5.148   PicBasic program for Example 1

Measuring the speed of the robot

The speed of the robot can easily be measured by moving it for a known amount of time and measuring the distance moved during this time. The speed is then given by

$$Speed = distance/time$$

In this project the robot moved forward for 10 s and the distance moved was 210 cm. Thus, the speed of the robot is $210/10 = 21$ cm/s.

Once we know the speed, we can move the robot forward or backward by any required amount. For example, to move the robot forward by 85 cm, the required time is approximately given by

$$\text{Time} = \text{distance/speed} = 85/21 = 4$$

Thus, the servomotors should be operated for 4 s. The value of loop-count M is then approximately given by

$$M = 4000/23 = 174$$

The required PicBasic code is

```
FOR J = 1 TO 174
PULSOUT 0, 170
PULSOUT 1, 130
PULSOUT 20
NEXT J
```

Turning left and right

Several techniques can be used to turn the robot left or right. One technique is to stop the servomotor on the side where we wish to turn. For example, we can turn right by stopping the right servo and turning the left servo anti-clockwise.

Another technique of turning a robot smoothly involves rotating both servos in the same direction and this is the technique we shall be using here. For example,

```
To turn RIGHT:
            Rotate left wheel anti-clockwise
            Rotate right wheel anti-clockwise

To turn LEFT:
            Rotate left wheel clockwise
            Rotate right wheel clockwise
```

The problem here is how many pulses to send to the servomotors so that the robot turns a complete 90° angle. This is something which can be found by trial and error.

The following PicBasic code rotates the robot right where the angle of rotation depends on variable R:

```
Turn_right:
      FOR J = 1 TO R
            PULSOUT 0, 170          ' Left wheel anti-clockwise
```

```
                    PULSOUT 1, 170              ' Right wheel anti-clockwise
                    PAUSE 20                    ' Wait 20 ms
          NEXT J
```

Similarly, the following code rotates the robot left where the angle of rotation depends on variable R:

```
    Turn_left:
          FOR J = 1 TO R
                    PULSOUT 0, 130             ' Left wheel clockwise
                    PULSOUT 1, 130             ' Right wheel clockwise
                    PAUSE 20                   ' Wait 20 ms
          NEXT J
```

It was found by experimentation that when R [r5] is equal to 13 the robot turns by about 90°. An example is given below.

Example 2

A mobile robot is controlled with two servomotors as shown in Figure 5.146, and a pen is attached to the front of the robot with the tip of the pen touching the floor. Write a PicBasic program which will move the robot as follows:

> Move the robot forward for 5 s
> Wait for 2 s
> Turn right
> Move the robot forward for 3 s
> Stop

Solution 2

The first action is to move the robot forward for 5 s. Thus, the value of M is

$$M = 5000/23 = 217$$

Then the robot is required to stop for 2 s and then turn right and move backward for 3 s. The value of M for this movement is

$$M = 3000/23 = 130$$

PicBasic program for this example is given in Figure 5.149.

```
'*********************************************************************
'
'                         ROBOT CONTROL
'                         ================
'
' In this project a mobile robot is controlled. The robot has two side wheels and
' a back caster wheel. Side wheels are connected to servomotors as follows:
'
'         Left wheel        RB0
'         Right wheel       RB1
'
' In this project the robot moves as follows:
'
'         Move the robot forward for 4 seconds
'         Wait for 2 seconds
'         Turn right
'         Move the robot forward for 3 seconds
'         Stop
'
' A PIC16F84 type microcontroller is used with a 4 MHz crystal
'
' Author:       Dogan Ibrahim
' Date:         January, 2005
' Compiler:     PicBasic
' File:         SERVO2.BAS
'
' Modifications
' ===========
'
'*********************************************************************

'
' Symbols
'
Symbol PORTB = $06                        ' PORTB address
Symbol TRISB = $86                        ' TRISB address
Symbol J = B0

        POKE TRISB, 0                     ' PORTB is output
```

Figure 5.149   (Continued)

```
'

' Move the robot forward for 4 seconds
'

        FOR J = 1 TO 217
                PULSOUT 0, 170
                PULSOUT 1, 130
                PAUSE 20
        NEXT J
'

' Wait for 2 seconds
'

        PAUSE 2000
'

' Turn right
'

        FOR J = 1 TO 13
                PULSOUT 0, 170
                PULSOUT 1, 170
                PAUSE 20
        NEXT J
'

' Move the robot forward for 3 seconds
'

        FOR J = 1 TO 130
                PULSOUT 0, 170
                PULSOUT 1, 130
                PAUSE 20
        NEXT J


        END                                     ' End of program
```

Figure 5.149   PicBasic program for Example 2

About the Companion Website

The Companion Website accompanying this book contains: the Demo version of the PicBasic Pro compiler, source files (.BAS) and object files (.HEX) of all the projects in the book, all the figures and the tables used in the book.

The files on the Companion Website are organised in the following folders:

DEMO	PicBasic Pro Demo application
PROJECT_SOURCES	Project source files (.BAS)
PROJECT_OBJECTS	Project object files (.HEX)
FIGURES	All the figures used in the book
TABLES	All the tables used in the book

This material is now available from the link below:

http://www.elsevierdirect.com/v2/companion.jsp?ISBN=9780750668798

Index

Note: Page numbers in italics refer to figures and tables.

Printed and bound by CPI Group (UK) Ltd, Croydon, CR0 4YY

03/10/2024

01040337-0001